UGLY BUGS
Nick Arnold and
Tony De Saulles
SCHOLASTIC

Scholastic Children's Books,
Euston House, 24 Eversholt Street,
London NW1 1DB, UK

A division of Scholastic Ltd
London ~ New York ~ Toronto ~ Sydney ~ Auckland
Mexico City ~ New Delhi ~ Hong Kong

First published in the UK by Scholastic Ltd, 1996
This abridged edition published by Scholastic Ltd, 2014

Index by Caroline Hamilton

ISBN 978 1407 14263 0

Printed and bound by CPI Group (UK) Ltd, Croydon, CR0 4YY

2 4 6 8 10 9 7 5 3 1

CONTENTS

Nick Arnold has been writing stories and books since he was a youngster, but never dreamt he'd find fame writing about Ugly Bugs. His research involved being stung, crawled over and covered in slime and he enjoyed every minute of it.

When he's not delving into Horrible Science, he spends his spare time eating pizza, riding his bike and thinking up corny jokes (though not all at the same time).

www.nickarnold-website.com

Tony De Saulles picked up his crayons when he was still in nappies and has been doodling ever since. He takes Horrible Science very seriously and even agreed to investigate what happens when your body is covered in leeches. Fortunately, his injuries weren't too serious. When he's not out with his sketchpad, Tony likes to write poetry and play squash, though he hasn't written any poetry about squash yet.

www.tonydesaulles.co.uk

INTRODUCTION

Science can be horribly mysterious. Not just science homework – it's a mystery how they expect you to do it all. No, I mean – science itself. For example, what do scientists do all day? Ask a scientist and you'll just get a load of scientific jargon.

*ENGLISH TRANSLATION. I'M LOOKING AT BEETLES THAT GLOW IN THE DARK.

It all sounds horribly confusing. And horribly boring. But it shouldn't be. You see, science isn't about all-knowing experts in white coats and laboratories and hi-tech gadgetry. Science is about us. How we live and what happens to us every day.

And the best bits of science are also the most horrible bits. That's what this book is about. Not science, but horrible science. Take ugly bugs for instance. You don't need to go very far to find them. Lift up any stone and something crawls out. Look into any dark, creepy corner and there's some ugly bug lurking there. Decide on a nice early-morning bath and you might discover you'll be sharing it with a huge hairy spider.

You see, ugly bugs bring science to life. Horrible life. Especially when you find out how a praying mantis catches its victims – and bites their heads off. Here's your chance to find out many more truly horrible facts about ugly bugs. And discover why for some ignorant adults an ugly bug – any ugly bug – is something to be swatted or sprayed out of existence.

Mind you, it's a good idea to keep this book out of reach of grown-ups because:

1 They might want to read it too.

2 It might give them bad dreams.

3 When you've read it you'll be far better informed than they are. You can tell them a few horrible but true scientific facts. And science will never seem the same again.

The worst thing about ugly bugs is that there are so many of them. There are thousands and thousands of different types. They have to be sorted out before we can even begin to get to know them. It's a horrible job – but someone has to do it.

Don't worry, though, it won't be you – here's a sorting method that scientists prepared earlier.

Each type of living thing is called a species and these species are put into larger groups called genera – a bit like belonging to a club. Groups of genera make families. Confused yet? You will be.

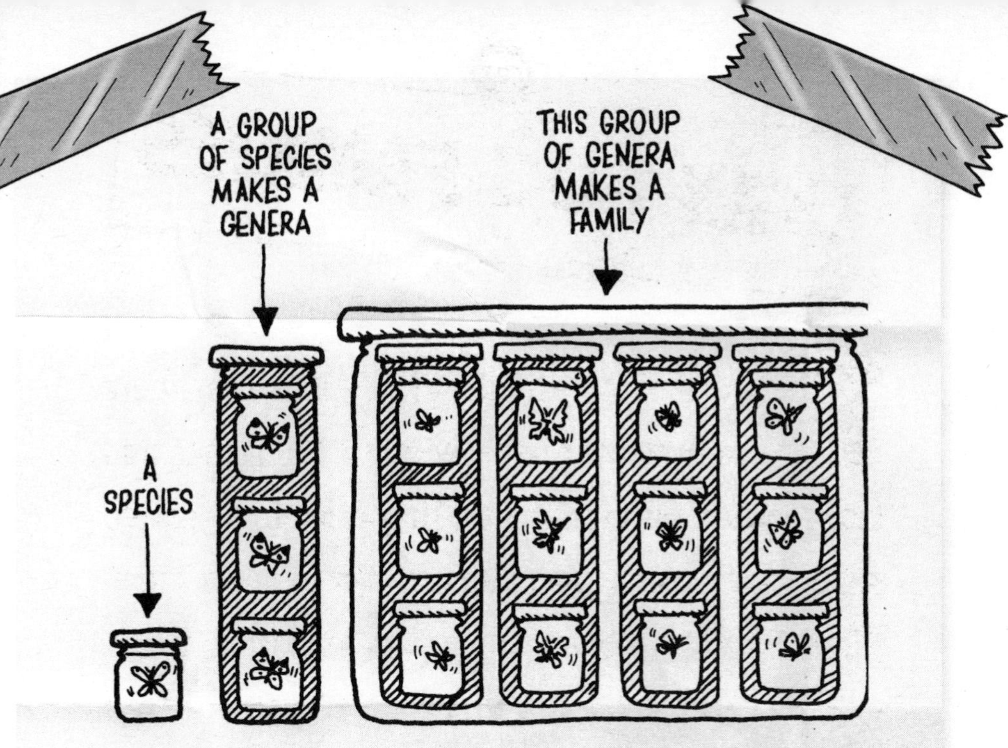

Like any family, ugly bug family members look a bit alike. But they don't all live together in a neat little home. If they did they might start fighting over who uses the bathroom first in the morning.

Groups of related families are known as "orders". And scientists lump orders together to make huge groups called "classes". (This is nothing to do with school, even if the classes have to follow orders.)

Here's an example of what we're talking about. This little bug is a seven-spot ladybird.

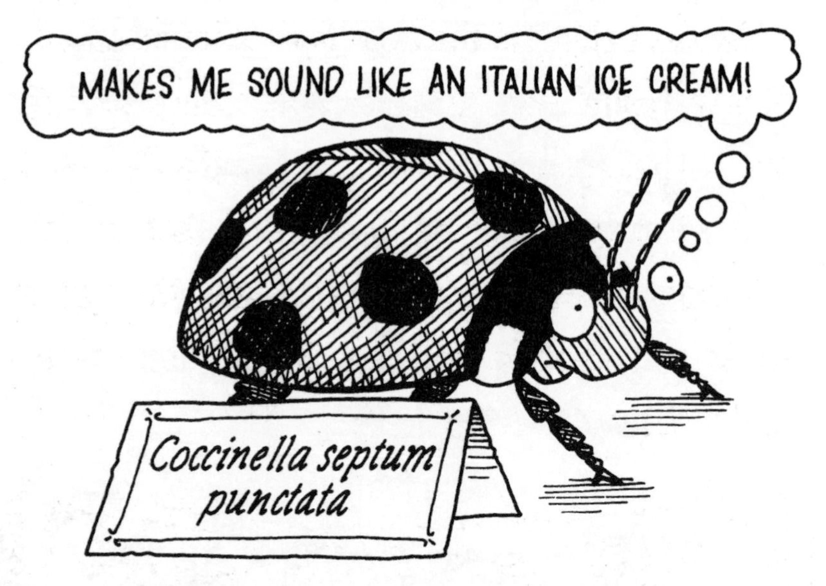

• Its scientific name is *Coccinella septum punctata* (try saying that with a mouthful of popcorn) – which is Latin for ... seven-spot ladybird.

• And ladybirds belong to an ugly bug family called *Coccinellidae* (cock-in-ell-id-day), or ladybirds. (Surprise, surprise!)

• Ladybirds belong to the order *Coleoptera* (coe-le-op-ter-ra) – that's beetles to you.

• Beetles belong to the class *Insecta*, or insects.

Simple really! And it makes good sense for ugly bugs to be organized. There are more than 350,000 species of beetle alone. Try sorting that lot into matchboxes! So, now you know how the system works, why not flip through the ugly bug family album? First let's meet some...

IRRITATING INSECTS

Insect bodies are divided into three parts – a head, a middle bit or thorax and a bit at the back called an abdomen. An insect has two feelers (antennae) on its head and three pairs of legs attached to its thorax. Scientists have identified about a million insect species with bodies like these and there are plenty more just waiting to be discovered.

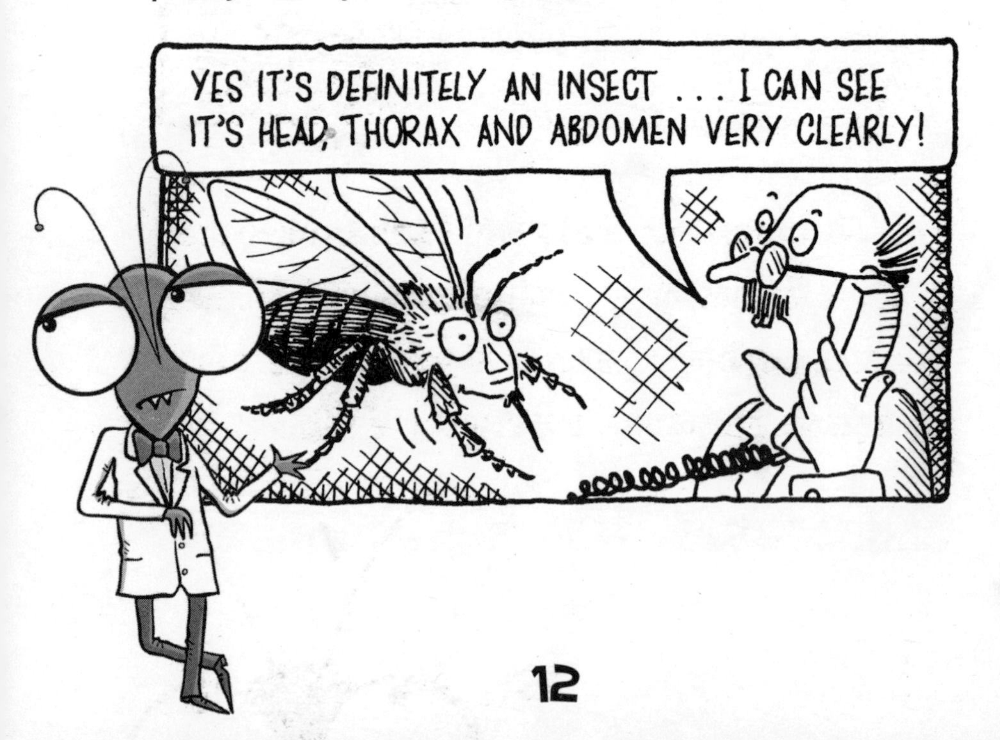

Earwigs At least 1,200 species. Earwigs get their name from the barmy belief that they crawl into your ears when you're asleep! They have mean-looking pincers at the back of their bodies. Males have curved pincers and females have straight ones.

Grasshoppers, crickets and locusts More than 13,000 species. They jump around and produce noises by rubbing their legs together to make themselves irresistible to the opposite sex.

Stick insects and leaf insects Over 3,000 species. Most live in tropical forests. Stick insects are so called because, well, they look like sticks, and leaf insects are so called because, you guessed it, they look like leaves. Either way they sit about all day looking like part of the furniture. Know anyone like that? It's a clever disguise, of course, but what a life!

Beetles At least 350,000 species in this order worldwide – that's more than any other type of animal. But you'd never be able to catch them all in a jam jar. Apart from their vast numbers, many of them are known only as a single example in a museum collection.

Termites More than 2,800 species. Termites like a nice hot climate. They are small soft insects but that doesn't mean they're a soft touch. Termites build nests that look like palaces and are ruled by kings and queens. Guard-termites are so serious about their work they sometimes explode in a bid to defend the nest!

Ants, bees and wasps Well over 120,000 species in this order worldwide. All members have a narrow waist between the thorax and the abdomen. Most have wings. (Worker ants don't develop wings – they're far too busy to go anywhere.)

Mantids and cockroaches At least 6,500 species. There's a strong family resemblance in their horrible habits. Cockroaches make midnight raids on the pantry. The praying mantis sits around cunningly disguised as part of a plant, and waits to pounce on its innocent victims.

Bugs Over 100,000 species in this order worldwide. They suck vegetable juices through straw-like mouths. Nothing ugly about that, you might think, except some do like a bit of blood now and then.

Flies Far more than 120,000 species in this order. They use one pair of wings for flying (which is what they do best). They also have the remnants of a second pair of wings that look like tiny drumsticks, and are actually used for balancing. Most irritating fly habit: flying backwards, sideways and forwards round your head. OK – so you know they're incredible fliers already. Nastiest fly habit: some types of fly like nothing better than to lick the top of a big smelly cowpat. And then pay a visit to whatever you were going to have for tea.

Sucking lice More than 500 species. Lice don't build their own homes. No. They live on other creatures. It's nice and warm there and you can suck a refreshing drop of blood whenever you feel like it. Lice can live on nearly every mammal – bats are one of the few exceptions. Or at least no one has ever found a louse on a bat.

Dragonflies, caddis flies, mayflies are three different orders totalling more than 17,000 species. They start off living in water and then take to the air. Traditional names for dragonflies include "horse stingers" and "devil's darning needles". Which is odd because they don't sting horses and you can't mend your socks with them.

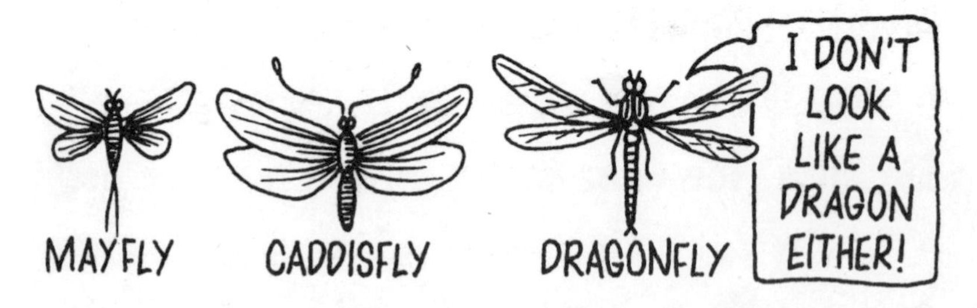

Butterflies and moths Well over 180,000 species in this order worldwide. They have two pairs of wings and their young start off as caterpillars. Then they hide in a case called a chrysalis and re-arrange their body parts before emerging as butterflies or moths. It's a bit like you spending a few weeks taking your body apart in a sleeping bag. And then putting it all back together in a different order.

So these are the ugly insects, but what about their even more repulsive relatives?

NASTY NON-INSECTS

If an ugly bug has got more than six legs – or no legs at all, it isn't an insect.

Slugs and snails Over 35,000 species on land and many live in the sea. Slimy slugs and snails belong to a huge group of animals called the molluscs that even includes octopuses. But slugs and snails are the only members of the group that have tentacles on their heads.

Centipedes and millipedes are two different classes of ugly bugs. There are about 2,800 species of centipede and more than 10,000 species of millipede. But sinister centipedes gobble up the poor little millipedes and *not* the other way round.

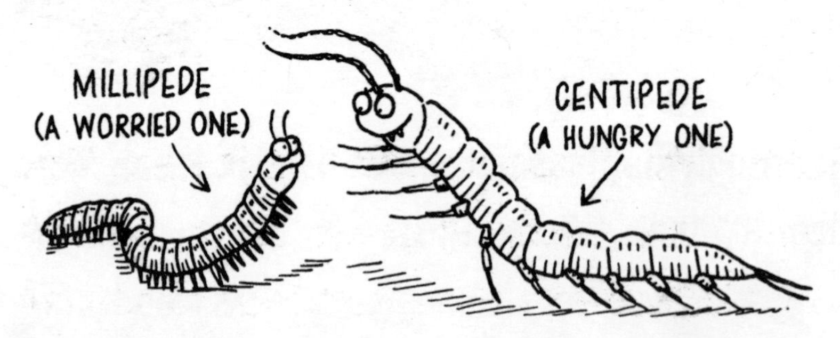

Woodlice Over 3,500 species. They all have seven pairs of legs. Woodlice, would you believe it, belong to the same class of creature as crabs and lobsters!

Spiders There are 37,000 species in this order but scientists think there may be up to five times that number waiting to be discovered! What a thought! Most spiders spin silken webs. They have eight legs, of course, and their bodies are divided into two parts.

Earthworms, bristleworms and leeches More than 16,000 species altogether. Leeches are the nasty bloodsuckers. When a leech sucks blood it can swell up to three times its original size. There are 300 different leech species. Yuck! One is enough!

Mites There are well over 45,000 species in this order. Unlike spiders, mites have a one-piece body. Many mites are under 1 mm long but they still have some hugely horrible habits. Some eat cheese rinds and the glue in old books. Others suck blood from animals.

So there you have it. Ugly bug families *are* horribly confusing. There are so many of them and they come in a horrible array of shapes and sizes. But they've got one vital feature in common – they're HUNGRY! Take the worms, for example, they like nothing better than a breakfast of slimy rotting leaves. And some worms have even more revolting tastes.

WEIRD WORMS

You can't get away from worms. They live in soil. Bet you didn't know their slimy relatives also live in the sea? You might also find them at the bottom of ponds and even inside other creatures. There are thousands of worm species with all sorts of ugly habits. But one thing they all have in common is that they're horribly weird.

WEIRD WORM VARIETIES

There are three main orders of worms. Flatworms, ribbon worms and segmented worms. So how can you tell which is which?

Weird flatworms

Surprisingly enough, flatworms get their name because they are pretty flat. Their bodies aren't divided into segments, and they're pretty slimy, too. They're probably the slimiest worm you'll come across.

For example, one type of flatworm, the parasitic tapeworm, can live inside an animal's stomach! Another, called dugesia, (dug-easi-er) picks on creatures smaller than itself and sucks them up. But if the tiddlers get a bit too big, ugly Dug wraps them up in a slimy parcel and just sucks bits off them.

Then there's the milky-white flatworm, a close relative of dugesia. It lives in water and it's almost see-through, so you can see what it ate for dinner. And when it wants to reproduce it sometimes tears itself in two!

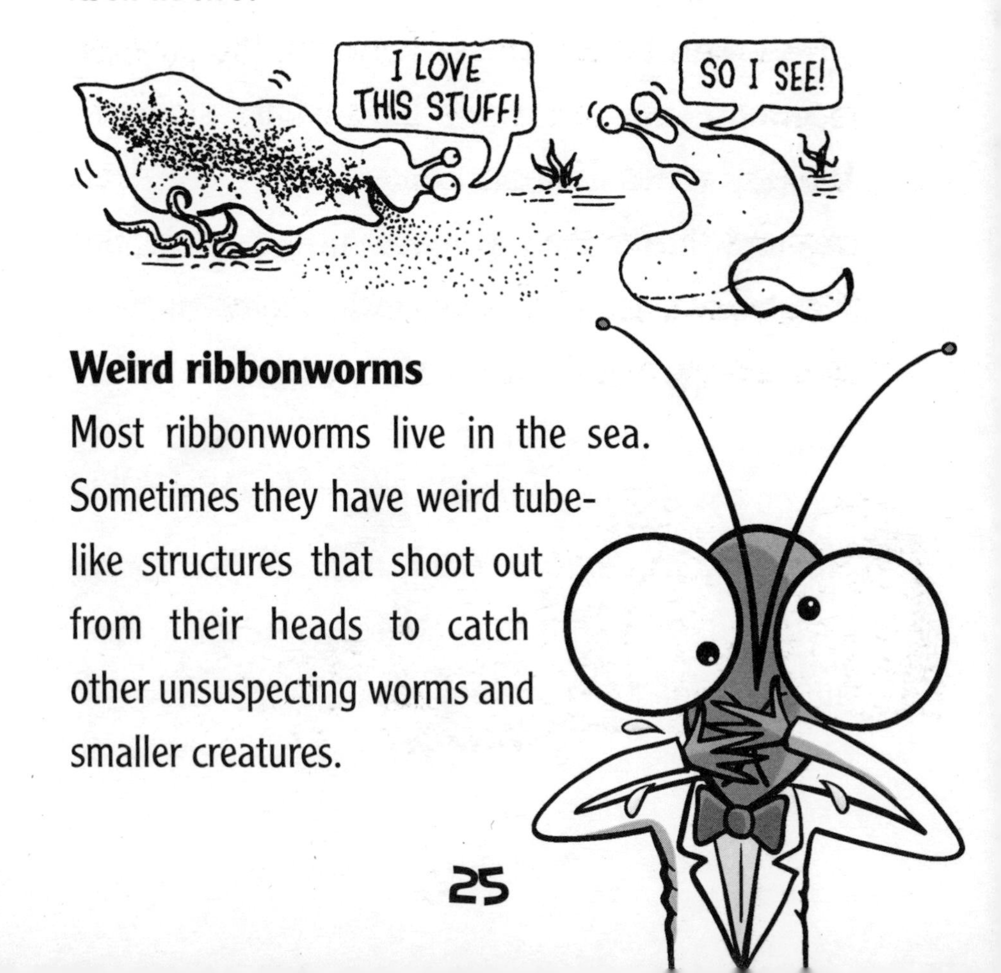

Weird ribbonworms

Most ribbonworms live in the sea. Sometimes they have weird tube-like structures that shoot out from their heads to catch other unsuspecting worms and smaller creatures.

Ribbonworms can be horribly long. The bootlace worm sometimes reaches several metres. Would you like to meet a worm that's as long as your bootlace?

Weird segmented worms

The weird worms in this gruesome group all have rounded bodies that divide into segments. Some of them are parasites, and can cause disease. Others might live in the soil, in the sea, or freshwater. They live on small plants and animals.

Bristleworms belong to this order. Maybe you've spotted them at the seaside? Some of them build tubes out of the sand and sit in them with their tentacles poking out. But uglier bristleworms crawl about looking for prey. They use their two pairs of jaws, two pairs of feelers and four tentacles to search out their food. They particularly enjoy sucking the insides out of snails. Yummy!

A sea mouse, on the other hand, has a mouse-shaped body that's all furry. Aahh, sounds quite cute, doesn't it? Except that this worm can grow up to 18 cm long and 7 cm wide. Sounds more like a sea rat!

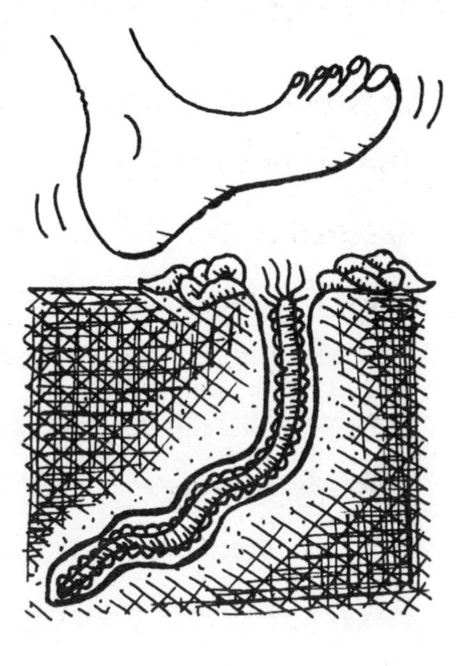

Want to get friendly with a member of the segmented worm family? Then let's get down to earth with earthworms...

Ugly bug fact file

Name of creature:	Earthworm
Where found:	Most soils worldwide
Distinguishing features:	Segmented body. See-through skin. Slides along by squeezing its body segments forward.

Visual identification

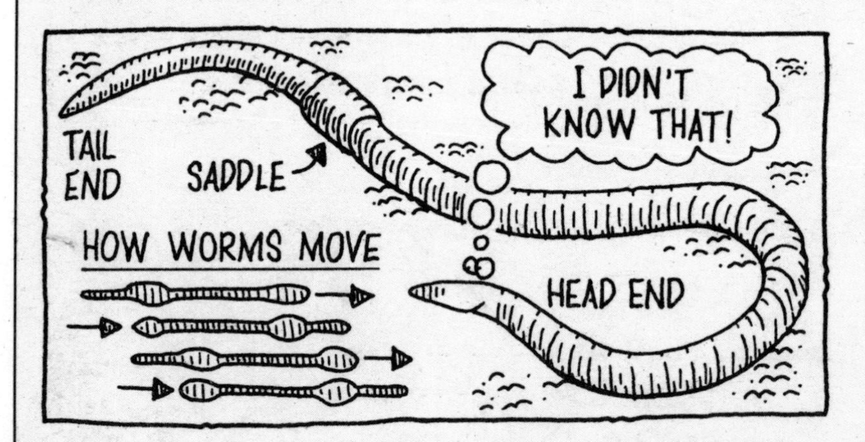

ARE EARTHWORMS AWFUL?

"Yes", according to people who don't like slimy wriggling creatures.

"No", according to some very famous naturalists.

In 1770 Gilbert White wrote that...

Charles Darwin liked earthworms too...

WHAT'S SO GREAT ABOUT THESE UGLY BUGS?

• Worm burrows mix up the soil bringing vital minerals to the surface so hungry plants can easily slurp them up.

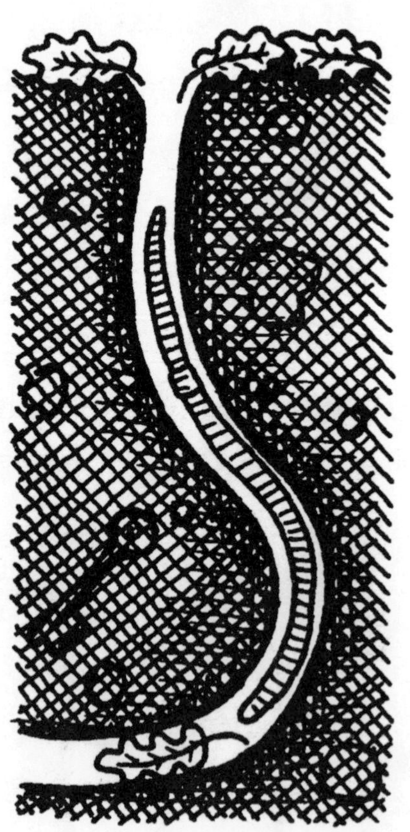

• Worm burrows make space for water and air to mix with the soil and reach plant roots.

• Earthworms drag leaves and other rotting material into their burrows. This rotting material can be taken up by plant roots.

So you see crops grow better in soil where there are lots of earthworms. In fact in Europe and the USA there are earthworm farms that produce up to 500,000 worms a day for sale to farmers. Good old earthworms!

But earthworms are still ugly bugs, so they do have some horrible habits. After the earth has passed through their bodies it ends up in ugly earthworm-shaped piles all over your beautiful front lawn. Earthworms love to guzzle lettuce and their burrowing can damage plant seedlings. Never mind – if your earthworms turn nasty you can always use them as fishing bait.

ARE YOU AN EARTHWORM EXPERT?

You may think that earthworms are deadly dull and boring. And of course, you'd be right. But delve a little deeper into their humdrum lives and you'll discover some slimy surprises. See if you can guess these answers.

1 How many worms could you count per hectare of farmland?
a) Three
b) 65,697
c) Two million

2 Why on earth do earthworms have bristles? (This is true. Just try stroking one – if you dare!)
a) To help them move along.
b) To stop the early bird from yanking them out of the soil.
c) To sweep their burrows clean.

3 How on earth does a worm accidentally bury a stone?
a) The stone rolls into a hole dug by the worm to catch beetles.
b) Worms push earth up from their burrows until the stone is covered.

c) Worms tunnel under the stone. The stone falls into the tunnel.

4 How long was the longest earthworm ever found?
a) 20 cm
b) 45.5 cm
c) 6.7 metres

5 Worms have a part of their body called a saddle. What on earth is it used for?
a) To give rides to earwigs.
b) To carry lumps of food.
c) To make an egg cocoon.

6 What happens when you cut a short piece off the end of a worm? (No need to try this out to discover the answer.)
a) It gets upset.
b) It grows a new tail.
c) It joins back together again.

7 What on earth do moles do to worms?
a) Eat them.
b) Bite their heads off.
c) Bite their heads off and let them escape.

ANSWERS

1 c) Amazingly enough. **2 a)** *and* **b)**! Trick question – sorry. **3 b)** This makes the soil level rise and things level with the soil sink down. **4 c)** It was a type of giant earthworm that lives in South Africa. This monster wriggled out of the ground in Transvaal in 1937. **5 c)** The saddle is a belt that moves the length of the worm, picking up the fertilized eggs. The worm wriggles free, leaving the eggs in a cocoon. **6 b)**. **7** Another trick question! The answer is all three! **a)** Moles love a juicy earthworm. **b)** When they're full they bite the worm's heads off. They put the worms in their "pantry" to enjoy later! **c)** When a mole bites a worm's tail off, it sometimes has time to grow a new tail and escape!

HOW TO CHARM A WORM

You will need:

- A fine day, but not too dry
- A lawn or flowerbed (make sure the soil is slightly damp)
- A pitchfork (optional)
- A hi-fi speaker (optional)

What you do:

1 You are going to pretend to be rain.

2 You can make vibrations by jumping up and down, playing music with the speaker facing the ground, or by sticking a pitchfork into the ground and wiggling it about a lot (this is also known as

"twanging"). Alternatively, use your imagination to create your own short, sharp shower. Anyone for a spot of tap dancing? But why does this make the worms come out?

Worms like rain because they have to keep their skin moist to prevent it drying out. When they feel rain drops hitting the ground they pop their heads out to take a look.

BET YOU NEVER KNEW

You can "charm" an earthworm. Every summer a primary school near Nantwich, England, hosts a weird competition. It's the world worm-charming championship. Yes – it's true. What a charming traditional pastime!

SLIMY SNAILS AND UGLY SLUGS

They're covered in slime, slide along very slowly and have eyes on the end of stalks. And if that's not ugly enough, they gobble up your garden lettuce. So it's not surprising that people don't like them. But are slugs and snails really that horrible? Do they deserve their rotten reputation? Yes they do. And here's why.

Ugly bug fact file

Name of creatures:	Slugs and snails
Where found:	Worldwide in the soil, in the sea and in fresh water. Land slugs and snails like damp places.
Distinguishing features:	Snails have shells on their backs. Slugs don't.

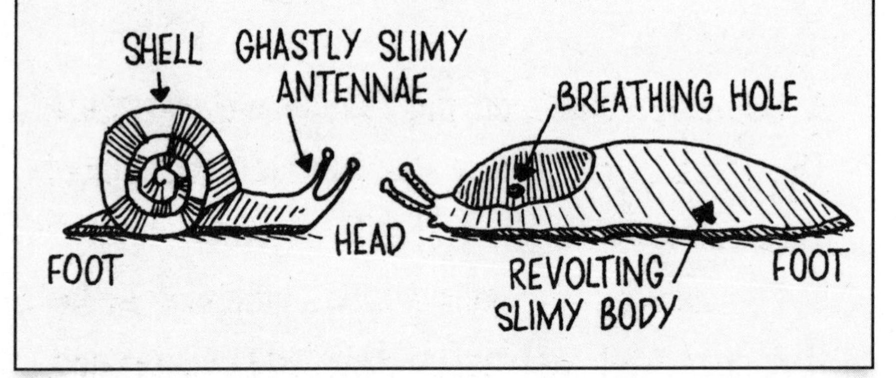

SEVEN SLIMY SNAIL FACTS YOU DIDN'T REALLY WANT TO KNOW

1 The largest snail in the world is the African Giant snail. It can be 34 cm from its shell top to its head! It eats bananas – and dead animals.

2 The garlic grass snail smells strongly of garlic.

OK – it's not really horrible. But it must give snail-eating birds horribly bad breath.

3 When a snail is chomping away on your mum's prize cauliflowers, it will be using its radula – that's its tongue. The radula is so rough it actually grates its food.

4 Some sea snails on the other hand, eat meat. These snails have a few sharp teeth – well suited for catching and chomping on their prey!

5 The slimiest sea snails are dog whelks. They lay their eggs in a tough capsule attached to the sea bed. But some of the youngsters seize and guzzle their own brothers and sisters as soon as they hatch out!

6 Another slimy sea snail is the oyster drill. Here's how an oyster drill drills:

a) It makes a chemical that softens up the oyster shell.

b) It scrapes the shell with its radula, repeating step **a)** as required.

c) It sticks its feeding tube through the hole and slurps up the juicy oyster!

7 But snails don't have it all their own way. A tiny worm lives inside the amber snail. Sometimes the worm releases chemicals that turn the snail's tentacles orange! This colourful display attracts a bird that nips off the snail's crowning glory. The worm begins a whole gruesome new life inside the bird. And the snail? It grows new tentacles. So that's all right then.

UGLY SLUGS

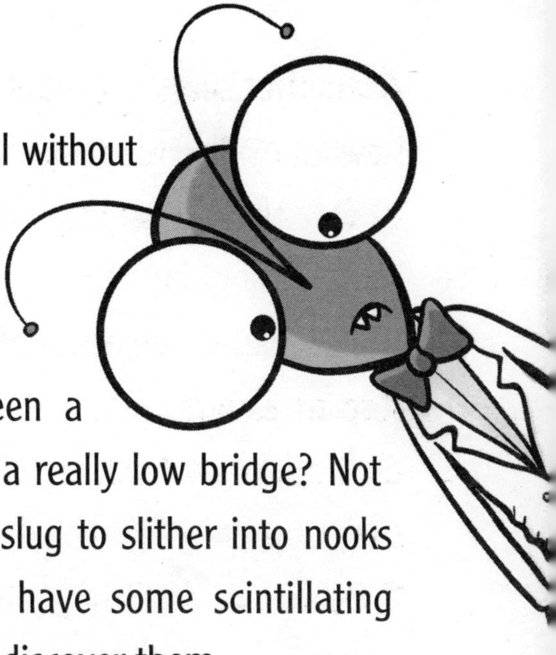

A slug is just a slimy snail without a mobile home on its back. Come to think of it – slugs have the right idea. Have you ever seen a snail trying to get under a really low bridge? Not having a shell helps the slug to slither into nooks and crannies. But slugs have some scintillating secrets. That's if you dare discover them.

DARE YOU MAKE FRIENDS WITH ... AN UGLY SLUG?

Here's how to snuggle up to a slug. Who knows, you could be in for a horribly interesting encounter!

1 First meet your slug. You can tell where there are slugs around by the horrible silvery slime trails they leave. They like to slither about in the open on warm damp summer evenings. So just follow a tempting trail until you find your slug lurking under the leaves of a small plant.

2 Enjoy that gooey, squelchy feel between your fingers as you put your slug in a glass jar.

3 Watch in amazement as your ugly slug climbs the slippery walls of the jar. It moves on a layer of slime produced by its foot. The sticky slime allows the slug to cling to the glass. Waves of movement push its foot forward. Think about it – could you climb up a glass wall on just one foot that's been dipped in something rather like raw egg?

4 Imagine you were a bird. Would you want to eat the slug? Not likely – the slime tastes disgusting! But hedgehogs think they are horribly delicious.

5 Put your new friend back where you found him/her. That way you'll stay friends.

If you go slug hunting in your garden on a warm damp night you might meet a shield-shelled slug. (Try saying that very fast – three times!) This sinister slug gets its name from a tiny shell on the top end of its body. But can you guess what it eats? Clue: It isn't lettuce.

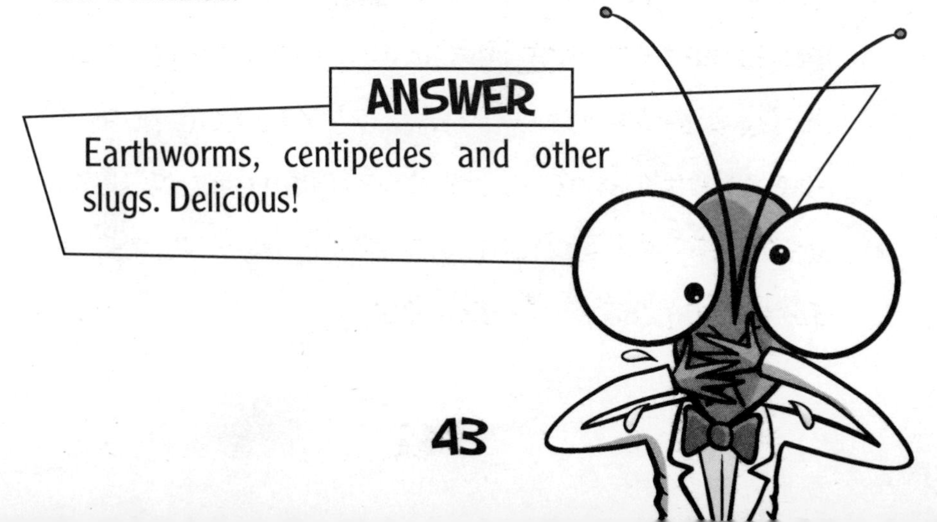

ANSWER

Earthworms, centipedes and other slugs. Delicious!

SEVEN UGLY SLUG FACTS

1 The largest British ugly slug is the great grey slug. It grows to 20 cm long!

2 But that's nothing! Some sea slugs are 40 cm long and weigh 7 kilos. They are also often brightly coloured.

3 And some of them have some horribly strange habits. Glaucus is a sea slug that floats upside down buoyed up by an air bubble inside its stomach.

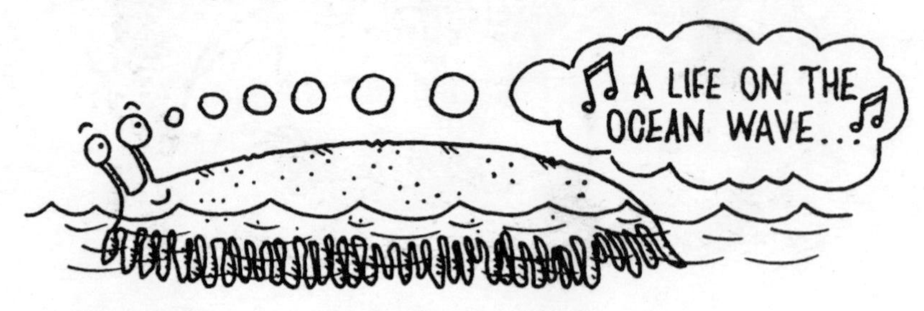

4 Meanwhile back on the farm, slugs and farmers are sworn enemies because ugly slugs eat or spoil crops. If slugs didn't eat potatoes there would be enough extra chips to feed 400,000 people for a year!

5 And land slugs have some horribly strange habits as well. Some ugly slugs can let themselves down from a height on a string of slime.

6 Like worms and snails, slugs are both male and female at the same time.

7 When slugs mate they cling together and cover themselves in slime. Then they fire little arrows called love darts at one another to get in the mood. Very romantic – if you're a slug!

BET YOU NEVER KNEW

An ugly slug can tell you which way the wind is blowing. It's true. A slug will always crawl away from the prevailing wind. Slugs do this to stop themselves drying out too quickly.

UNDERWATER UGLIES

Why not relax by a peaceful pond or river and forget about horrible ugly bugs? Some chance! Ugly bugs like water even more than you do. And those murky waters hide some pretty ugly undercurrents.

Imagine a pond as a kind of living soup. It's full of tiny plants and animals. The largest animals are always trying to eat the smaller ones, and the smaller animals are trying to eat even smaller animals, and they're all trying not to be eaten by each other. Scientists call this a food web because you get in a tangle if you try to figure out who eats who.

A pond is a perilous place to live. And its not just other animals that bugs have to watch out for. There are plenty of hazards all year round.

And at all times, horrible humans chuck in harmful rubbish and poisonous pollution. And then they drain the pond!

UGLY UNDERWATER LIFESTYLES

Every freshwater ugly bug has developed its own methods of living and eating. See if you can match each ugly bug to its loathsome lifestyle.

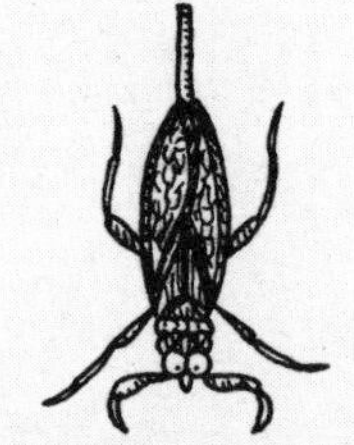

a) WATER SCORPION

1 Hangs under the water surface and breathes through a tube. Grabs a passing bug in its claws and sucks out the juices.

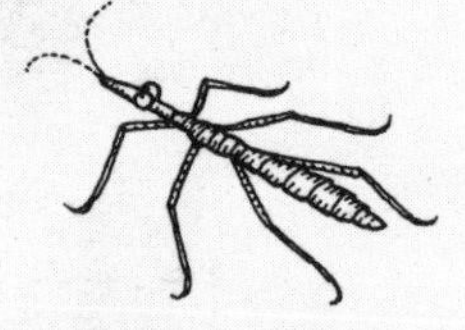

b) WATER MEASURER

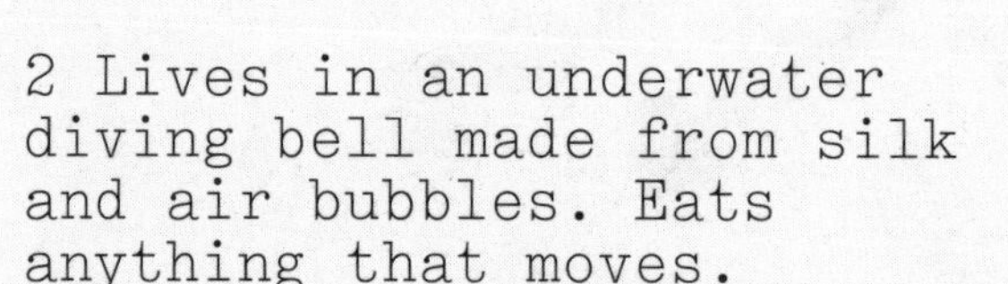

2 Lives in an underwater diving bell made from silk and air bubbles. Eats anything that moves.

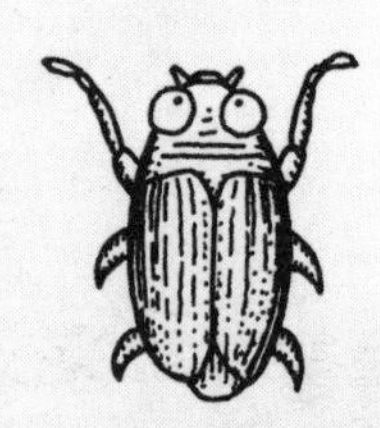

c) WHIRLIGIG BEETLE

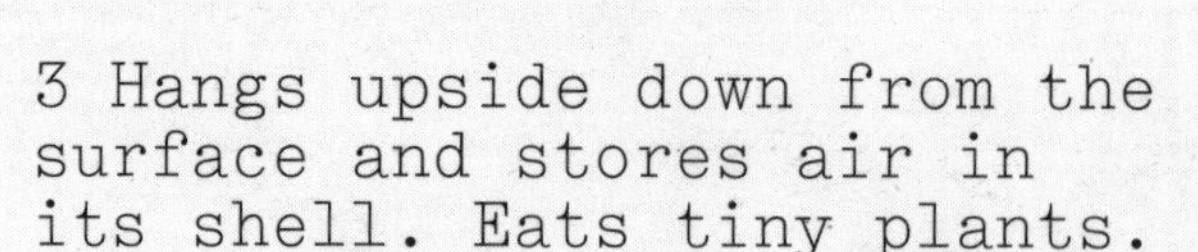

3 Hangs upside down from the surface and stores air in its shell. Eats tiny plants.

4 Walks around on the surface looking for bugs that have fallen in. Its light body and widely spaced legs ensure it doesn't break

d) GREAT POND SNAIL

the water tension. (That's the springy top surface of the water.)

e) WATER FLEA

5 Lives in the water and leaps to escape. Lives off tiny plants.

f) WATER SPIDER

6 Swims round in circles on the surface and dives to escape danger. Has four eyes – one pair of eyes above the water and one pair below. It can also fly! Eats other pond bugs.

ANSWERS

1 a) 2 f) 3 d) 4 b) 5 e) 6 c)

UGLY WATER SPORTS

As long as conditions in the pond are right and there is plenty of food, life for a pond ugly bug must seem one long holiday. Is this a holiday you could do without?

WELCOME TO UGLY BUG WATER WORLD!

The water sports centre where leisure is lethal!

The small print. We can't guarantee your safety. If you get eaten it's not our fault ~ OK?

A Great Dive!

Dive into danger with the great diving beetle. Store air bubbles under your wings to stay down for longer. Also learn to grab and guzzle any underwater edibles.

Rafting & fishing

Enjoy a lazy paddle with our resident raft spider. As you float by on your leaf raft try a spot of fishing. Just dip one of your eight legs in the water to attract little fish.

Power Beetle Boat racing

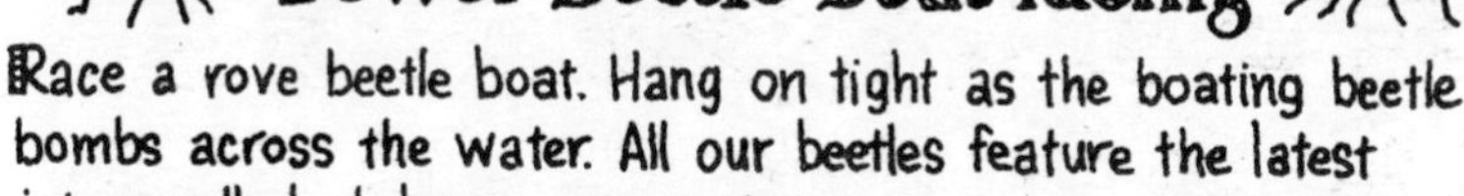

Race a rove beetle boat. Hang on tight as the boating beetle bombs across the water. All our beetles feature the latest jet-propelled abdomen gas engines.

Water good swim!

Learn basic backstroke with our brilliant backswimming water boatman beetle. Swimming on your front lesson taught by his assistant – the lesser water boatman.

Now you've worked up an appetite. Where better to relax than our exclusive underwater eating places?

The Caddis Fly Larva Cafe

Gravel built with silk wallpaper – it's the perfect place for a relaxing and informal meal. Book now before your caddis fly chef grows up and flies away. Vegetarian? Don't worry! The nearby Veggie Caddis Fly Cafe offers a choice of tiny slimy plants and bits of rotting leaves. Warning to patrons. Beware the treacherous trout. They sometimes try to eat the cafe.

Cafe

BEWARE OF THE TROUT!

BET YOU NEVER KNEW!

May or June is when mayflies hatch out and have the day of their lives. Literally. They only live a day or so. They mate, lay their eggs and die. Ugly bugs and fearsome fish go crazy banqueting on the bodies.

LOATHSOME LEECHES

Lurking at the bottom of your local pond or canal is a creature that makes the others seem quite likeable. There's no way of disguising it. These creatures are *loathsome*!

Ugly bug fact file

Name of creature:	Leech
Where found:	Worldwide in water or damp rain forests.
Horrible habits:	Sucks blood.
Any helpful habits:	Used in medicine to ... suck blood (surprisingly enough)!
Distinguishing features:	Long segmented body with suckers at the back and front.

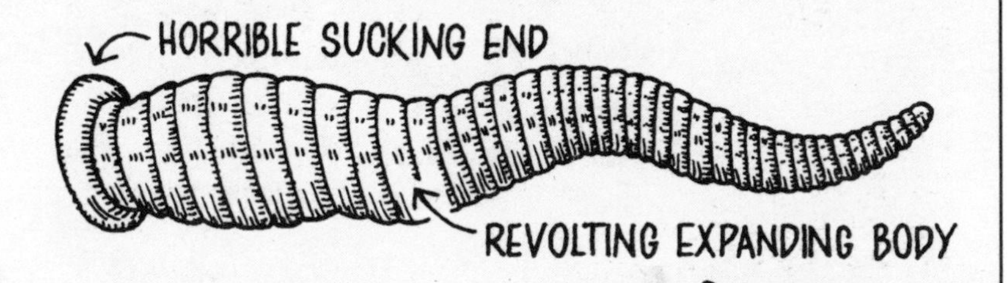

THE MOST LOATHSOME LEECH AWARDS

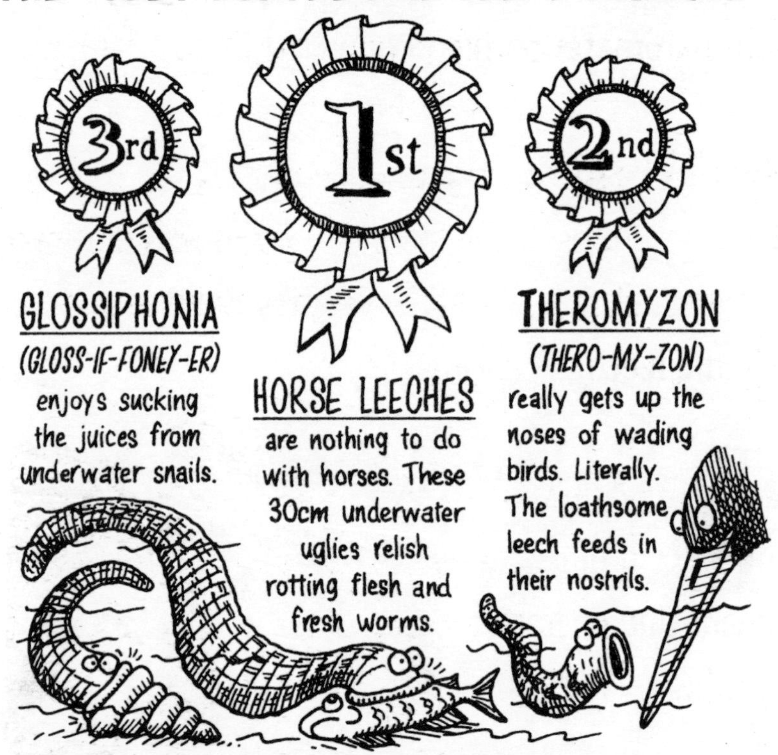

A LOATHSOME LEECH BAROMETER

But even leeches have their uses. Here is a vile Victorian invention it's best *not* to try. Simply place a leech in a jar of fresh pond water. Cover the top of the jar with a fine cloth and secure tightly. Feed your barometer on blood now and then.

How to read the barometer

1 Leech climbs to the top of the jar means that rain is expected. If the weather settles down again, so will the leech.

2 Lazy leech lies on the bottom of its jar means fine or frosty weather.

3 Restless leech shows that a storm is on its way.

Who hasn't looked under a stone at one time or another and seen an assortment of horrible-looking creatures? Chances are that these creepy-crawlies included centipedes, millipedes and woodlice. Now you might think that because these creatures live in the same place they'd all be mates. Well, you'd be horribly wrong. Centipedes like to eat millipedes – when they get the chance. And that's just the start of their disgusting differences!

ANTENNAE
HEAD
CENTIPEDE
MILLIPEDE

Ugly bug fact file

Name of creatures:	Centipedes and millipedes
Where found:	Worldwide, often amongst leaf litter and rotten wood.
Distinguishing features:	Centipede: Segmented, slightly flattened body. Two jointed legs on each segment; two long feelers. Millipede: Segmented, rounded body. Four jointed legs on each segment; two short feelers.

CREEPY COMPARISONS

1 Feet count Millipede means "thousand feet" – which just goes to show that some scientists can't count. Most millipedes have between 80 and 400 feet. Centipede means "hundred feet". But once again the scientists got it horribly wrong! Some centipedes have only 30 feet.

2 Walking When a millipede walks, waves of movement pass up its body so that it glides along. When a centipede walks it raises alternate legs just as you normally do. It has extra long legs at the back so it doesn't trip up.

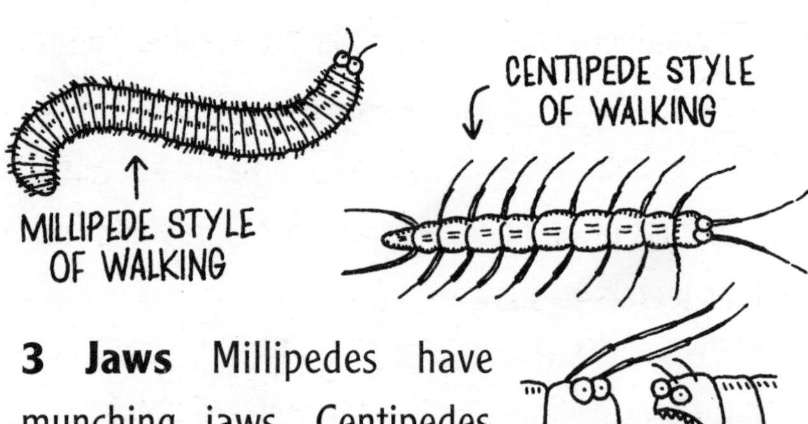

3 Jaws Millipedes have munching jaws. Centipedes have poison fangs. They're both pretty ugly.

4 Romantic problems Millipedes have a big problem – they can't see very well. So male millipedes have developed some strange ways of attracting a mate.

- Some bang their heads on the ground.
- Others let out a loud screech.

- Some produce special scents.
- Others rub their legs together to make sounds.

A male centipede, on the other hand, has other things on his mind. All centipedes are horribly aggressive and the female he fancies is quite capable of eating him! So first of all he walks around her, tapping her with his feelers to show he's friendly.

MURDEROUS MILLIPEDES AND CENTIPEDES

Centipedes enjoy eating millipedes – when they get the chance. But the millipedes often put up a fight!

Here's what happens... Centipede attack plan: Spear prey on fangs and inject poison. Once prey stops wriggling – nibble at leisure.

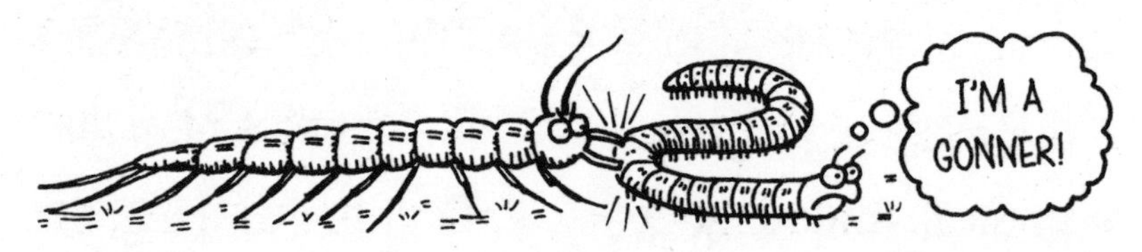

Millipede defence plan: Curl up in a ball. Squirt nasty fluid from stink glands on sides of its body.

Who do you think has the best chance of winning – the menacing millipedes or the sinister centipedes?

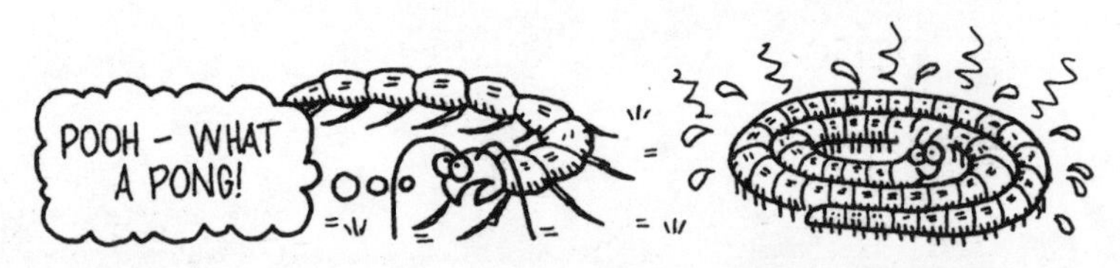

In some parts of the world, centipedes and millipedes can grow to gigantic proportions. Giant millipedes can measure up to 26 cm long.

Some of these monsters have fearsome fangs.

One type of centipede in the Solomon Islands has a particularly painful bite. People have been known to plunge their hands into *boiling water* to take their minds off the pain! In Malaysia the local centipede's bite has been described by travellers as worse than a snake's. And in India there are even scarier stories of people who were *killed* by giant centipede bites.

Mind you, the millipedes aren't much better. In Haiti in the West Indies giant millipedes attack the local chickens and sometimes blind them with jets of poison! Other giant millipedes produce little puffs of poison gas. The gas kills any attacking animal.

But size doesn't save either the giant centipedes or the giant millipedes from a horribly gruesome death. In the African savannah giant hornbills are often seen plodding along looking at the ground. Suddenly they will nab a passing centipede in their long beak, and the centipede has no chance to bite the bird back. Scrunch, crunch, gobble and poor old deadly giant centipede has turned into another scrumptious snack for the hornbill.

Other centipedes get carried away by armies of ants. OK, the centipede can easily kill a few hundred ants but when it's 10,000 ants to one centipede, the poor old centipede doesn't stand a chance!

Giant millipedes have it tough too. Grey meerkats often feed on millipedes. Funny thing is that the meerkats always screw their faces up in disgust when they're eating. Well, who would expect a millipede to taste good?

DARE YOU MAKE FRIENDS WITH ... A MILLIPEDE?

Now for the good news. In the UK millipedes are quite harmless. Just as long as you handle them gently and as long as you don't try to make a meal of them. Here's how to make a meal for them instead, just to show what a good pal you are.

1 First catch your millipede. (And make sure it is a millipede, not a centipede!) Millipedes lurk in shady places, so try looking under leaf litter, compost or loose tree bark.

2 Pop your new friend into a small jar partly filled with earth – and a piece of a bark so it can hide.

3 Then serve up a tasty treat. A millipede's mouth would water at the thought of a ripe raspberry, a piece of potato skin, a mouldy old lettuce leaf or a little chunk of apple.

4 Place the jar in a dark secluded place.

5 Next day find out which delicious dish the millipede preferred.

6 Then it's time to say goodbye to your millipede mate. So pop your guest back where you found it. There is sure to be plenty of food and shelter there and let's hope there are no centipedes loitering nearby. Otherwise your millipede will end up on someone else's menu!

A WOODLOUSY LIFE

Along with the millipedes and centipedes, at the bottom of your garden live hundreds – no *thousands* – of woodlice. There are 50 different species of woodlice in Britain and they're all shy and nervous so make sure you read this book *quietly*. The most common species are the imaginatively named common woodlouse and the pill bug – not to be swallowed for a headache.

Ugly bug fact file

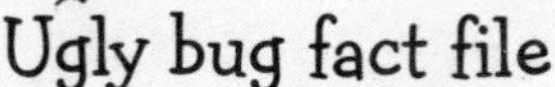

Name of creature:	Woodlouse
Where found:	Worldwide in damp, dark places where there is rotting material, e.g. slimy brown leaves.
Distinguishing features:	About 15 mm long with seven pairs of jointed legs and two feelers. Segmented armour around its body allowing it to move easily.

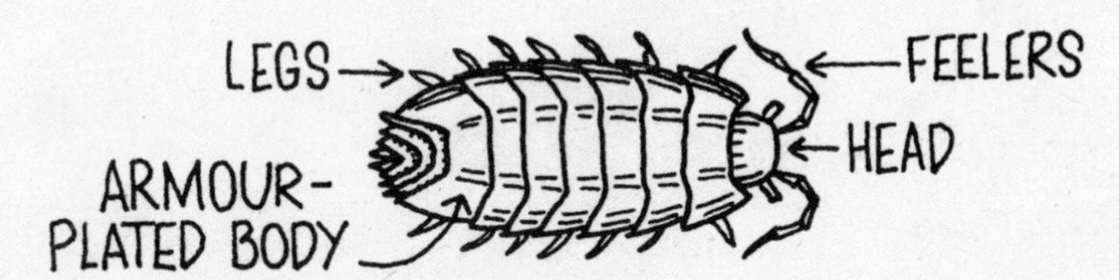

A pill bug can roll itself into a ball (but please don't try bouncing one) – the common woodlouse can't do this. Some people think woodlice are boring. But, as always, they are wrong. Woodlice are horribly interesting.

TEN TERRIBLY INTERESTING FACTS ABOUT WOODLICE

1 Not a lot of people know this, but a woodlouse is not a louse! In fact, country people call woodlice some extremely un-boring names.

2 Woodlice have extremely interesting relatives. Crabs, shrimps, prawns, lobsters and woodlice are all part of the crustacean family. Many people are extremely interested in eating their seaside relatives. Not many people are interested in eating woodlice you might think ... but you'd be wrong.

3 This is not a horrible habit but a delicious delicacy. Salted and fried woodlice are an African speciality. They're eaten like crisps!

4 Woodlice themselves have a horribly boring diet, though. They prefer bits of rotting plants and moulds. It's not everyone's cup of tea. But somebody's got to eat it, otherwise we'd be knee deep in the stuff. And woodlice do liven up their diet with the odd interesting dish ... like other woodlice for instance. Or their own droppings and their skin after they've shed it.

5 Woodlice start off as eggs in their mum's tummy pouch. Four weeks later they hatch as tiny woodlice. Baby woodlice live with their parents, which is an interesting way for an ugly bug to start life because most insect eggs are abandoned by their mothers. It's terrible, but true!

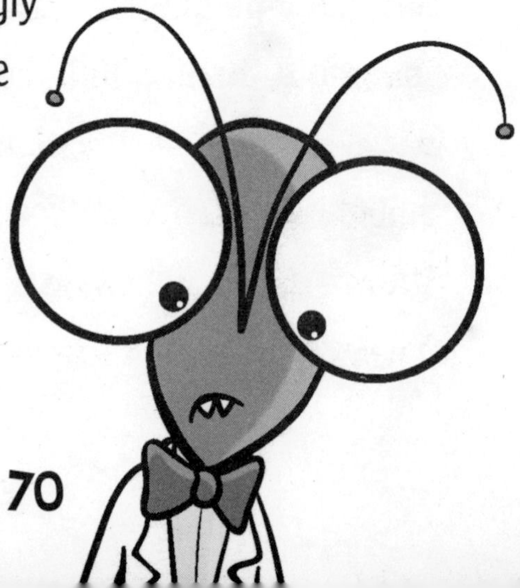

6 And woodlice lives are full of drama and excitement. They put most TV soaps to shame. Yes. Woodlice never go to bed early with a mug of cocoa. They get to sleep all day and go out every night. And then they break into your home.

7 You're most likely to see woodlice in wet weather, because the biggest danger for a woodlouse is drying out. Interestingly, every year millions of baby woodlice come to a sad and sticky end by simply shrivelling up.

8 Some woodlice live in horribly interesting places. One variety lives inside yellow ants' nests and eats their droppings. Another type of woodlouse lives by the seaside under piles of slippery rotting seaweed.

9 Woodlice have some interesting if deadly enemies. The most dangerous of these is the dreaded woodlouse spider. Once grabbed in the spider's pincer-like grip, a woodlouse is doomed. The spider injects its poison and the woodlouse dies in ... seven seconds. Quite an interesting way to go.

10 And there are some horribly interesting woodlice pests. Such as the tiny worms that sometimes live inside them ... and kill them. Or the disgusting fly larvae that creep into a woodlouse's body and eat it from the inside out.

DARE YOU MAKE FRIENDS WITH ...

A WOODLOUSE?

Woodlice may not be the masterminds of the ugly bug world, but they've learnt a trick or two about how to survive. So why not put your woodlouse to the test? Make a note of what it does, then try to work out for yourself what makes a woodlouse tick.

1 First, find your woodlouse under a stone or a log, or in a damp comer.

2 Get a piece of wood (like a ruler) and try and get your woodlouse to climb onto it at different angles. Does your woodlouse:

a) walk off in the other direction

b) easily climb onto the wood

c) struggle to get onto the wood?

3 Get a box with half the lid cut off. Find out which half the woodlouse likes best: **a)** light **b)** shade.

4 Tip your woodlouse onto a tabletop and poke it gently with the point of a pencil. This is a pretty scary thing to do to a woodlouse (it'll be scary for you, too, if you do it on the dinner table – at dinnertime). Does your woodlouse:

a) roll up in a ball

b) run away

c) clamp down on the ground

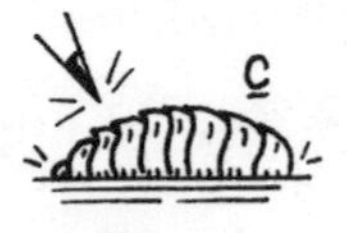

d) pretend to be dead

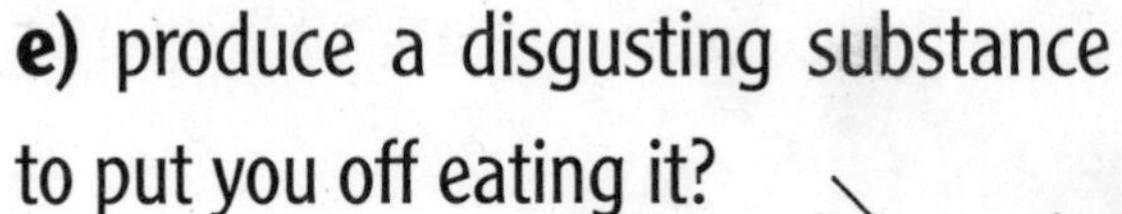

e) produce a disgusting substance to put you off eating it?

5 Don't forget to pop your woodlouse back unharmed where you found it.

Did you discover ... that your woodlouse could easily climb out of danger ... it sheltered in the shade, so as not to dry out in the sun ... it had various sneaky survival tricks when it sensed it was in danger of attack?

With such a collection of tricks up its many trouser legs you'd think we'd be even more overrun by woodlice than we are. Well, we would be – if it wasn't for competition from a group of bugs so ugly that they make the woodlouse seem cuddly. Enter the Insect Invaders!

INSECT INVADERS

Seen from any point of view insects are a horribly important group of ugly bugs. Insects are the most varied, the most ruthless, the hungriest and according to some people the most disgusting life form on the planet. There may be over 30 million varieties of insect. That's TEN times more than all the other types of animal *put together*.

Not surprisingly, you can find insects virtually anywhere you look. That's if you really want to look! It's also not surprising that they have a big effect on our lives. And it's mainly as invaders – of crops, homes, schools... Nowhere is safe from the insect invaders!

INSECT BITS AND PIECES

Despite their many differences, insects have the same basic features. We've caught this cute little beetle so that you can have a close look at it...

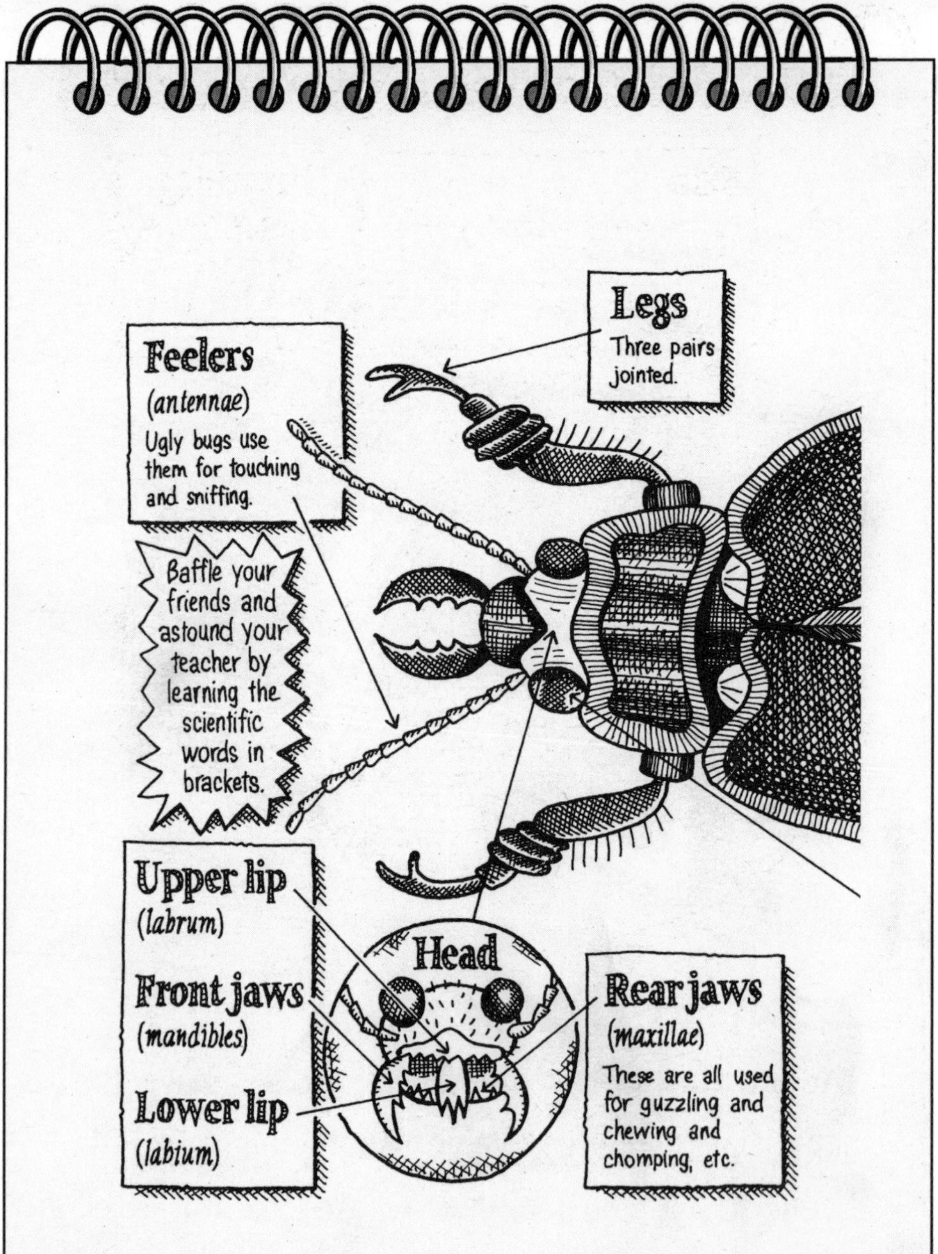

Legs

Three pairs jointed.

Feelers

(antennae)

Ugly bugs use them for touching and sniffing.

Baffle your friends and astound your teacher by learning the scientific words in brackets.

Upper lip

(labrum)

Front jaws

(mandibles)

Lower lip

(labium)

Head

Rear jaws

(maxillae)

These are all used for guzzling and chewing and chomping, etc.

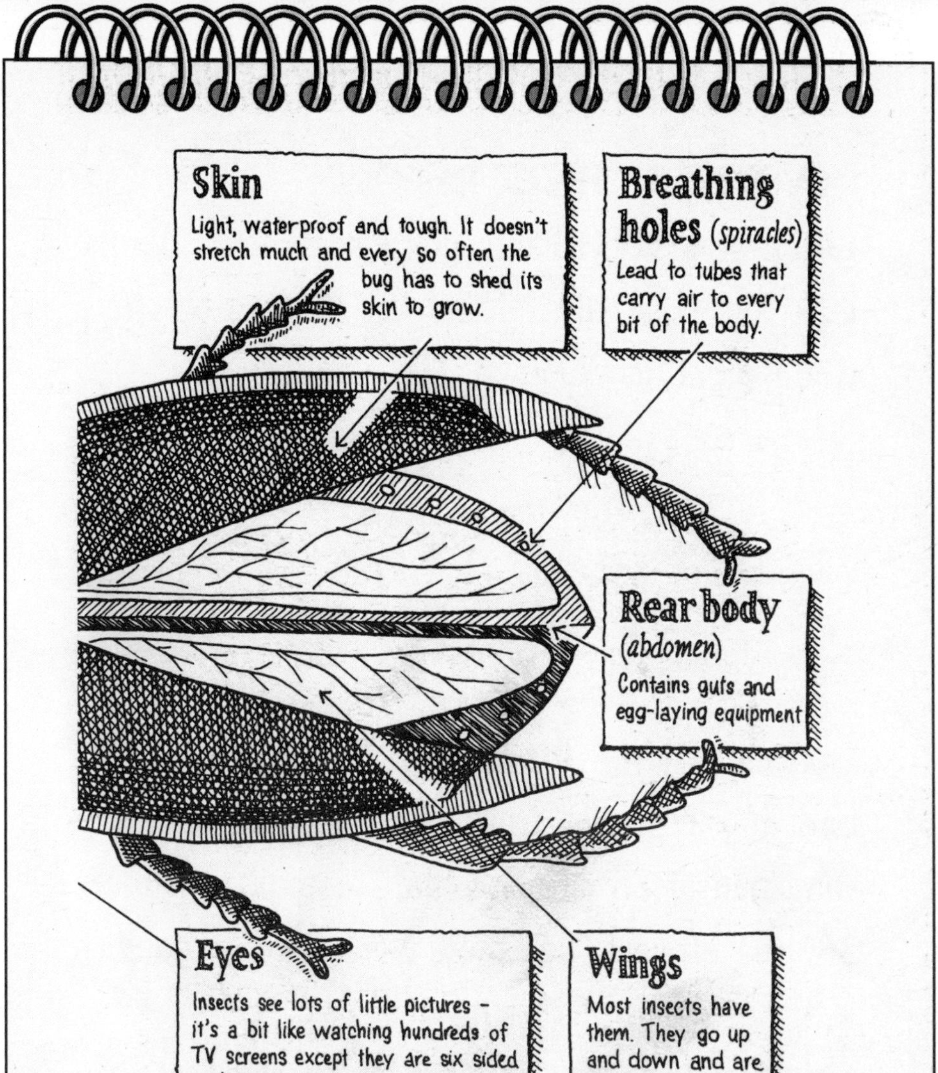

Skin

Light, waterproof and tough. It doesn't stretch much and every so often the bug has to shed its skin to grow.

Breathing holes *(spiracles)*

Lead to tubes that carry air to every bit of the body.

Rear body *(abdomen)*

Contains guts and egg-laying equipment

Eyes

Insects see lots of little pictures – it's a bit like watching hundreds of TV screens except they are six sided and none gives a good picture. But they are good for spotting anything that moves and is worth eating!

Wings

Most insects have them. They go up and down and are controlled by the muscles inside the body.

REVOLTING INSECT RECORDS

Longest insect Giant stick insects from Borneo look like ugly old sticks. And they grow to a whopping 33 cm long.

Largest flying insect The Queen Alexandra's birdwing butterfly from New Guinea boasts a wingspan of 28 cm. But that's nothing – a mere 300 million years ago there were giant dragonflies with wingspans of 75 cm!

Smallest insects Cute little fairy flies are actually tiny wasps only 0.21 mm long. The good news is that they don't sting humans.

Heaviest insect A single Goliath beetle from central Africa can weigh up to 100 grams.

Lightest insect The lightest insect is a species of parasitic wasp. It would take 25 million of them to weigh as much as one Goliath beetle!

Fastest flying insect There's a species of Australian dragonfly that can reach 58 km (36 miles) per hour.

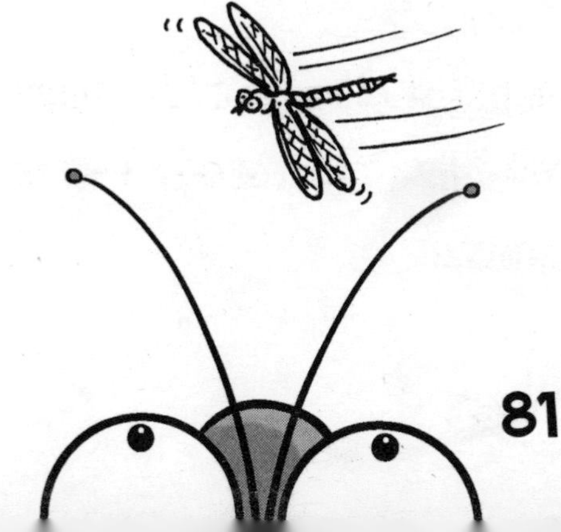

Fastest breeding insects Aphid females usually give birth to live young. Inside these are developing bugs. Inside the developing bugs there are more developing bugs, and so on. One female aphid can produce millions of descendants in a single summer.

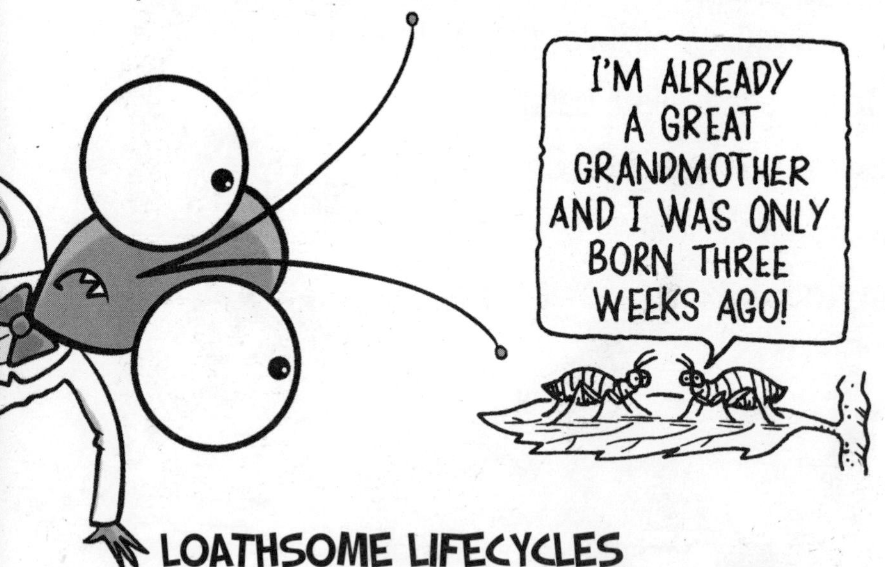

LOATHSOME LIFECYCLES

Some ugly bugs only change a bit as they grow up and some change completely – so there are two types of horrible insect lives.

Lifecycle 1

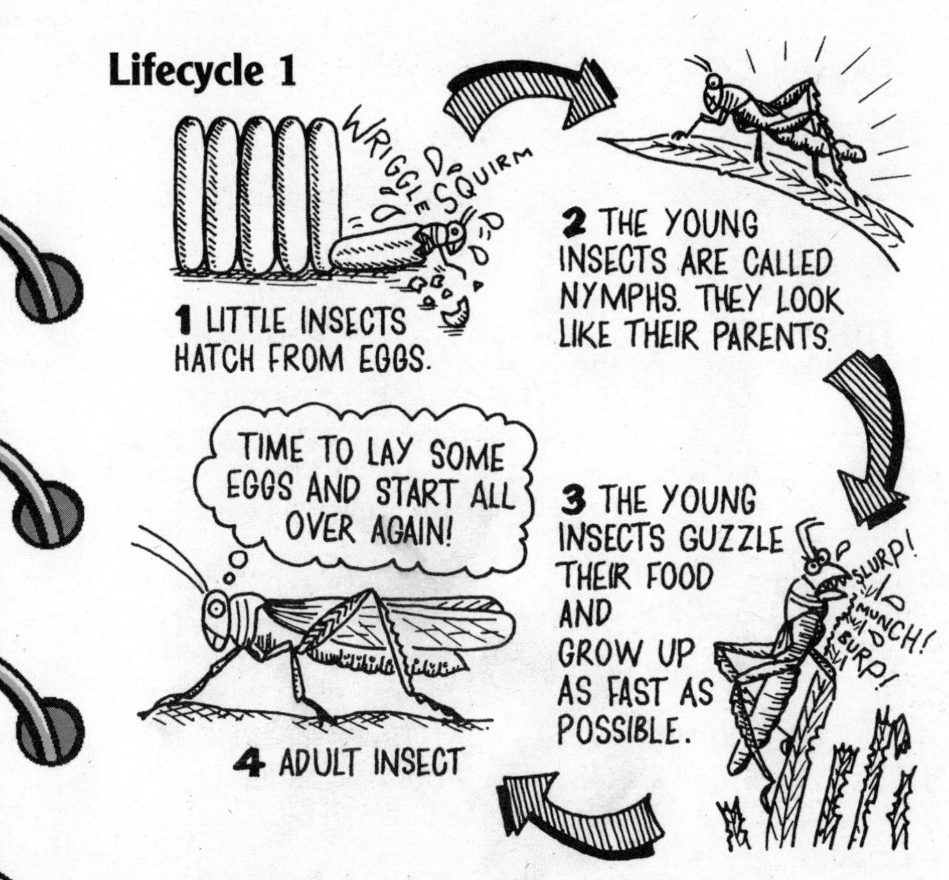

The scientific name for this loathsome lifecycle is "incomplete metamorphosis" (met-a-more-foe-sis). This describes a changing body. Mantids, locusts, dragonflies develop like this.

Lifecycle 2

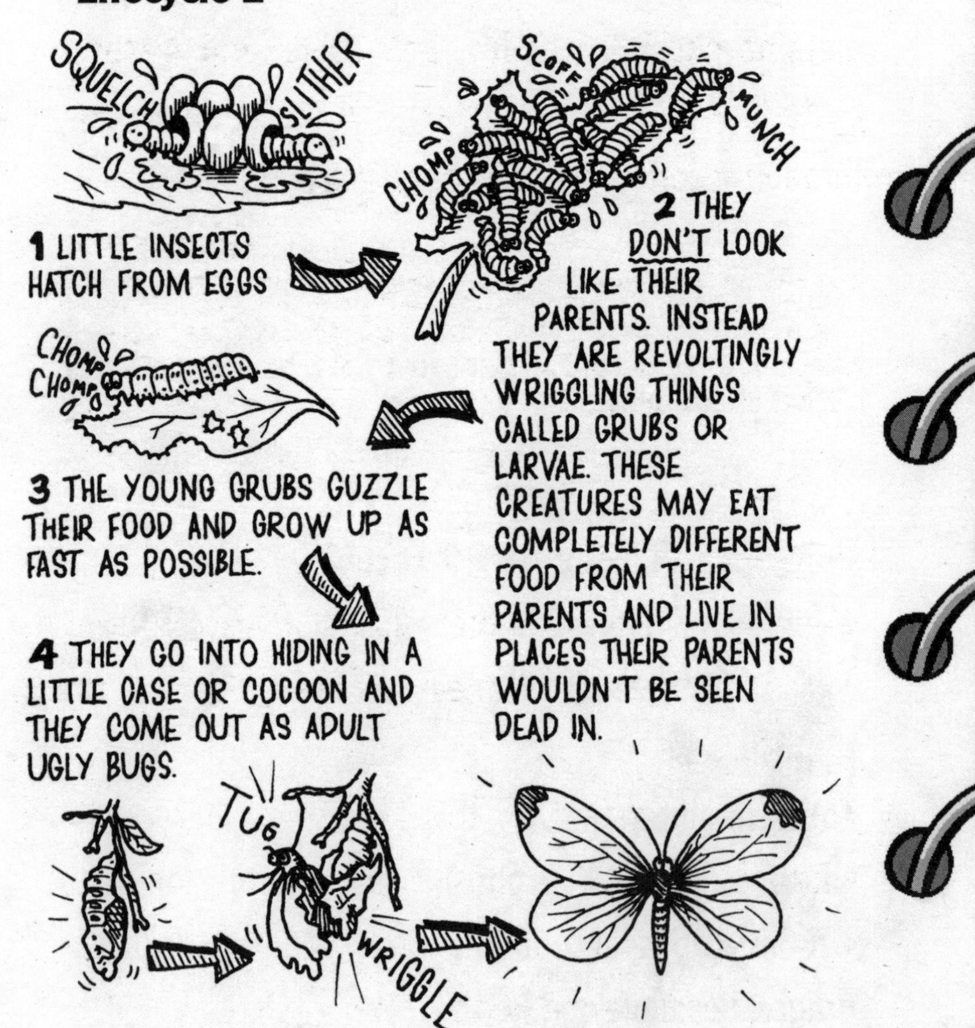
SQUELCH
SLITHER
SCOFF
MUNCH
CHOMP
1 LITTLE INSECTS HATCH FROM EGGS
2 THEY DON'T LOOK LIKE THEIR PARENTS. INSTEAD THEY ARE REVOLTINGLY WRIGGLING THINGS CALLED GRUBS OR LARVAE. THESE CREATURES MAY EAT COMPLETELY DIFFERENT FOOD FROM THEIR PARENTS AND LIVE IN PLACES THEIR PARENTS WOULDN'T BE SEEN DEAD IN.
CHOMP
CHOMP
3 THE YOUNG GRUBS GUZZLE THEIR FOOD AND GROW UP AS FAST AS POSSIBLE.
4 THEY GO INTO HIDING IN A LITTLE CASE OR COCOON AND THEY COME OUT AS ADULT UGLY BUGS.
TUG
WRIGGLE

The scientific name for this lifecycle is "complete metamorphosis". Beetles, ants, bees and wasps, butterflies and moths, flies and mosquitoes go through a complete metamorphosis.

BET YOU NEVER KNEW

People used to believe that insects such as flies developed from rotting meat and dead animal bodies. What a nice thought!

TERRIBLE TABLE MANNERS

Would you like to go to dinner with an insect? If so, you'd better learn how to eat like one.

You will need:

- A new sponge
- Tape
- A drinking straw
- A saucer of orange juice

What you do:

1 Cut a small piece from the sponge.

2 Tape it to the end of the drinking straw.

3 Try to suck up a saucer of orange juice.

Congratulations! You're eating like a fly. Flies also sick up digestive juice. It helps them to dissolve their food before they slurp it up! (Don't try this!)

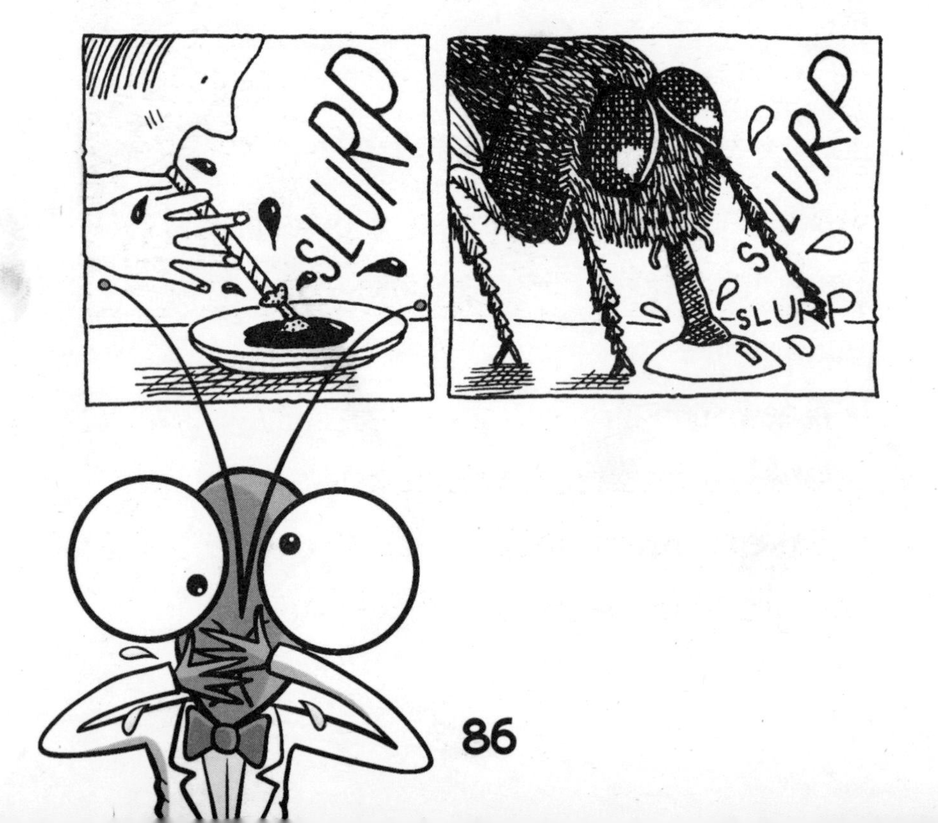

TOO HORRIBLE TO WATCH

Films are full of insects – especially scary films. There are giant ants and giant flies. And it's amazing how many space monsters look like insects.

In fact, film designers often study ugly bugs to get good ideas for a really ugly monster.

But who needs made-up insect monsters when some real-life insects are far more creepy?

First prize for creepiness Diopsid flies can see round corners because their eyes are on long stalks.

Second prize for creepiness There's a type of weevil that has a neck as long as the rest of its body. And no one knows why it's so long!

HORRIBLE BEETLES

Most people think that beetles look horribly ugly. Especially big black beetles that run over your foot and seem to enjoy it. The bad news is that of all the many orders of insects, beetles are the biggest group. And it's getting bigger because scientists are always discovering new species! Amazingly enough there is only one basic design for a beetle body.

Ugly bug fact file

Name of creature:	Beetle
Where found:	Worldwide. Found just about anywhere you can imagine except in the sea, although some beetles live on beaches.
Distinguishing features:	Most beetles have short feelers. Folded forewings over the hind wings protect the beastly beetle body.

UNBELIEVABLE BEETLES

With so many species of beetle it's inevitable that some of them are horribly amazing. And some of them have unbelievably horrible effects on human homes and food. But which of these beetles are too unbelievable to be true?

TRUE OR FALSE?

1 The biscuit beetle eats, would you believe, biscuits. That's the bad news. The good news is that it doesn't like chocolate biscuits – only those nasty digestives you don't eat anyway. True/false

2 The cigarette beetle eats (howls of amazement) cigarettes. Its larvae especially like the tobacco and they never take any notice of the health warnings. True/false

3 The violin beetle doesn't eat violins – it just looks like a violin with legs. It lives amongst layers of fungus in trees in Indonesia. True/false

4 The ice-cream beetle used to live in the Arctic where it ate small flies. More recently it has become a pest of cold stores where its favourite food is tutti-frutti ice cream. True/false

5 "Tippling Tommy" is the nickname for a beetle that bores holes in wine and rum barrels. Tippling Tommy is actually a teetotaller. That's to say it never touches the alcohol inside the barrels – it prefers the wood! True/false

6 The drug store beetle is the name given to a biscuit beetle that lives in medicine cabinets. It enjoys slurping up some medicines, including many poisons! True/false

7 The giant gargling beetle is a rainforest beetle that takes a mouthful of dew and makes a loud gargling sound first thing in the morning. True/false

8 The bacon beetle beats you to breakfast every time by looting your larder in the night and munching your meats. Its favourite food is – you guessed it ... bacon! True/false

9 The museum beetle is so fond of living in the past that it lives in dusty old display cases and eats museum specimens. Its favourite food ... preserved ugly bugs. True/false

10 Deathwatch beetles live in wood. Some English churches contain families of beetles that have lived there for hundreds of years. True/false

ANSWERS

1 True. **2** True. But they are more usually found on tobacco plants. **3** True. **4** False. Even beetles can't survive extreme cold. **5** True. **6** True. It is especially fond of traditional medicines made from dried plants. **7** False. Beetles don't use mouth-wash. **8** True. But you could always fry the beetle instead of the bacon. **9** True. The visitors have to view the living beetles instead. **10** True.

DARE YOU MAKE FRIENDS WITH ... A LADYBIRD?

One kind of beetle that definitely does exist is the ladybird. If you've ever wanted to get to know one socially this is your opportunity.

1 First look for some tempting aphids. They can be white, brown or black "greenfly" which you'll find on your rose bushes and other plants in summer.

2 Break off a small branch or leaves swarming with aphids and place the lot in a jam jar.

3 Add a ladybird. You can find them from the late spring onwards on bushes and fences. Watch your ladybird get to work. Lovely ladybirds can gobble up 100 greenfly a day.

4 Handle your ladybird gently and let it go after lunch. Do you really want to know what happens if things go wrong and your date gets upset? Try tickling it gently with a leaf of grass. It will produce nasty tasting liquid. This will definitely put you off eating it. If you tickle it more it will roll on its back and pretend to be dead – a quick way to end your lunch date. If you upset it a lot it will bite. And beware – they *do* bite!

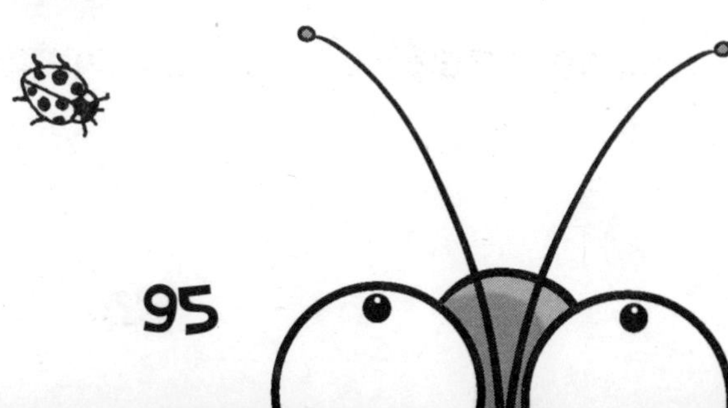

HOW NOT TO UPSET A LADYBIRD

During lunch you can discuss any subject with the ladybird without causing offence. This is because ladybirds don't understand English. Silly rhymes such as...

...will not offend your ladybird in the least. This is also because:

1 Ladybirds don't have homes. A sheltered leaf is good enough for them. So it is unlikely they would be bothered if their home was on fire.
2 Ladybirds can fly but no ladybird would ever fly towards a fire. (Only maniac moths do that.)
3 Ladybirds don't give two hoots for their children. Once their eggs are laid, that's that!

GOT A TOUGH JOB? GET A BEETLE TO DO IT

Beetles don't only come in a horrible variety of shapes and sizes. They also have a mind-boggling array of lifestyles, and where there's a job to be done there's a beetle at the ready.

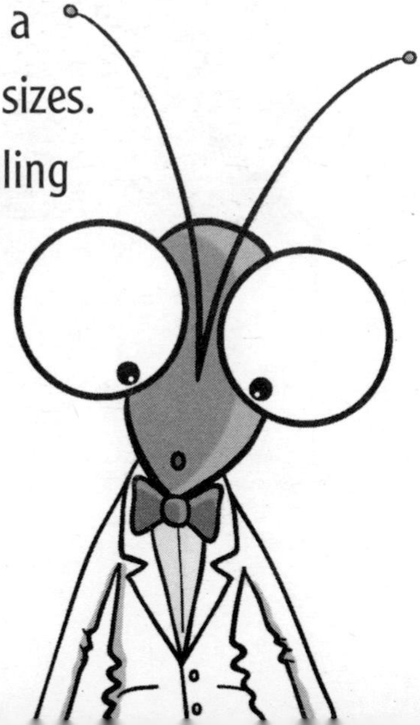

BEWARE IT'S A BOMBARDIER!

Get yourself the ultimate in personal self-defence systems! Beat off the bullies with a bombardier beetle gun. Unique self-mixing action for nasty boiling chemicals. Amazing internal heating system in abdomen heats chemicals to temperature of 100°C and fires at 500 to 1000 squirts a second!

The bombardier beetle gun is maintenance free. Just let it crunch on a few smaller insects now and then.

ELM BARK BEETLE TREE SURGEON

Unsightly elm trees getting you down? Need a bit more light? Call us now. Try our unique Dutch elm disease fungus formula – a revolting little rootless plant that terminates trees. We'll soon get in under the bark and wipe out the woody weeds!

- Forest felled.
- No job too large.

Disease established in UK – 1970s. Over 25 million elms eliminated.

BRIGHTEN UP YOUR HOME

With a firefly lantern. As used in Brazil, the West Indies and Far East. Firefly lanterns cast a soft green or yellow light from the bodies of female fireflies. Forty fireflies are as bright as one candle. They need no power or batteries – it's all done with chemicals by your friendly firefly.

Sexton Beetle
and Sons and Daughters

Dead? Just call in your friendly family funeral directors. No job too large. We'll bury anything even if it means ten hour shifts. Free personal limb chopping service to make burials easier. Professional after-care service. Our little grubs will look after the grave. No fee charged but they do like to come to the funeral feast. That's to feast on the dead body of course!

NEED ANY DUNG SHIFTED?

Scarab Beetle Services will get rid of the lot. Dung ball rolling and burying our speciality. What's more we'll even lay our eggs on it and get our grubs to eat it!

'Scarab beetles were round before the dung hit the ground They had 7,000 on the job and soon got rid of it all! My savannah has never looked tidier.' A.N.Elephant, Africa.

JEWELLERY WITH A MIND OF ITS OWN

Ever wanted some jewellery that puts itself away at night? Buy some living jewel beetle jewellery as worn in many parts of the world. Choice of beautiful metallic colours including gold. Breaks the ice at parties, e.g. 'And what would your earring like to eat?' Manufacturer's warning: Don't allow your jewellery to lay eggs on your furniture. The grubs can lunch on your lounge suite for up to 47 years before turning into more jewel beetles.

BEETLE BATTLES

Beetles don't have much of a family life, but they take proper care of their property – if they don't, they'll soon find themselves in big trouble.

STAG BEETLE WRESTLING

If you were a male stag beetle, this is how you would defend your territory (your territory would be a bit of a tree branch). The aim of the game is to drop your opponent off the branch...

You will need:

• A pair of giant jaws that look like deer antlers

What you do:

1 Eye up your opponent.

2 Grab him round the middle with your jagged jaws and try to flip him on to his back – that's easier said than done when he's trying to do the same to you...

3 If you lose, *you* fall off the branch and land on your back, where you risk being dissected and chewed by a waiting bunch of ... awesome ants.

AWESOME ANTS

Everyone knows about ants. They're easy enough to identify in the summer when they march into your home to inspect your kitchen. Ants can be pretty awful – they get everywhere from your plants to your pants – but they can be awesome, too, in all sorts of horrible ways.

Ugly bug fact file

Name of creature:	Ant
Where found:	Worldwide on land. They always live in nests.
Distinguishing features:	Most ants are less than 1 cm long. Narrow waist between thorax and abdomen. Angled feelers.

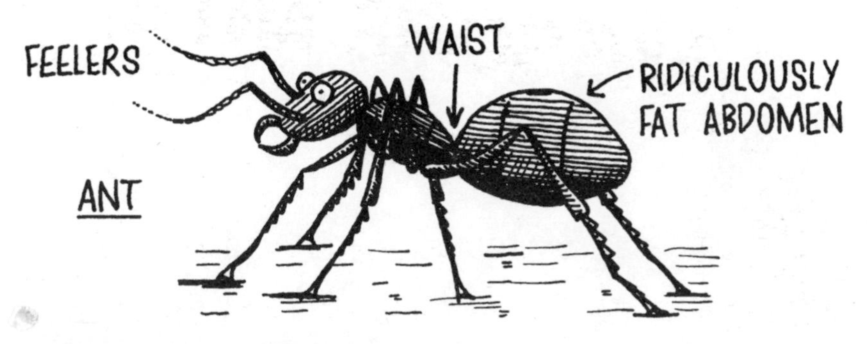

AWESOME ANT ANTICS

1 Since 1880 German law has protected red ants' nests from destruction. Why? Because the ants from each nest eat an awesome 100,000 caterpillars and other ugly pests every day.

2 Honeypot ants squeeze sticky honeydew from aphids. They're doing the aphids a favour – they don't need the sickly stuff. The ants keep feeding this honeydew to particular ants in their nest to make them swell up like little beads. The swollen ants then sick up the honeydew to feed the rest of the nest. Awful!

3 Weaver ants make their own tents from leaves sewn together with silk. Their larvae produce the silk and the awesome ants use their young as living shuttles weaving them backwards and forwards! The adult ants just have to touch their larvae with their feelers whenever they need a bit more silk.

4 South American trapjaw ants have huge long jaws. (Well, they're huge by ant standards.) They catch little jumping insects called springtails in their jaws and then inject them with poison.

But what's really awesome about these ants is that they also carry their eggs, or larvae, in those

gruesome jaws as gently as any mother carries her baby – isn't that nice!

5 Leaf-cutter ants grow their own crops. The ants cut up the vegetation and mix it up with their droppings to make fertilizer. Then they grow fungus on it for food. They even weed unwanted kinds of fungus from their garden and put it on their compost heap. When a leaf-cutter queen leaves to start a new nest she always takes a bit of fungus with her to start a new garden.

6 And after the hard work of farming comes the harvest. Harvester ants live in the desert where they gather seed grains and make bread by chewing it all up and removing the husks. The ants store their bread until they get hungry.

7 The Australian bulldog ant has an awfully ugly bite. Not only is the bite painful, but this appalling ant then injects poison into the wound! Thirty stings can kill a human in 15 minutes. This is probably the most dangerous ant in the world...

8 Or is it? In the jungles of Africa and South America lurks something even more awesome. It's 100 metres long and 2 metres wide. It eats anything foolish enough to get in its way. It reduces lizards, snakes and even larger animals to skeletons. And even big strong humans run for their lives rather than face it. Nothing can fight against it and live. What is this terrifying creature? Is it an ant? Well yes, actually, it's a column of 20 million army ants. The ants have no settled home. They spend their time invading places and being awesomely awful to any creature that gets in their way. If you live in South America it could get rid of cockroaches in your home, but you'd have to get yourself out of the way first.

9 Red South American Amazon ants fight fierce battles against their deadly enemies – the black ants. Red ant foot patrols are sent out to find a way into the enemy nest. They leave a trail for the main army to follow. The main army attacks and the Amazon ants use their curved jaws to slice off the heads of the opposing black ants. Some of the Amazon ants spray gases to further confuse the black ants. Then the Amazon ants retreat with their prisoners – the black ant grubs.

The grubs quickly pick up the smell of the Amazon ants and this fools them into thinking they're red ants too! But they aren't, and the poor befuddled black ants spend the rest of their lives as slaves to the awesome Amazon ants.

10 Marauder ants in Indonesia even build their own roads. These roads are often as long as 90 metres –

and if you're ant-sized, that's *awesome*. Some of the roads even have soil roofs to protect them. And the ants have to follow a strict highway code:

A Always keep to your own part of the road. Returning ants in the middle, outwards ants at the edges.

B Move anything that gets in your way. If it's big, gnaw it. If it's small, get the younger ants to carry it off the road. If it's edible, bring it back to the nest (100 workers can shift one earthworm, 30 workers can shift one seed.)

C If you cross any other ant roads . . . kill the other ants. All ugly bugs that get in the way must be eaten alive.

BET YOU NEVER KNEW

There are 10,000 species of ant. But they do have some things in common.

- **An ant nest is ruled a queen who spends her life laying eggs.**
- **All the ordinary worker ants are female.**
- **Males only hatch out at mating time and they die once they have mated!**

An intellig-ant man

Almost as awesome as the ants themselves, are some of the humans who studied them. Take Baron Lubbock, for example...

Baron Lubbock (1834-1913) was an expert on everything. He wrote more than 25 books, and over 100 scientific reports.

And those were just his hobbies. In politics he introduced . . .
. . . Bank holidays to Britain.
Hurrah!
He was an artist, too. Lubbock drew pictures for one of Charles Darwin's books on . . .
Barnacles
He travelled around Europe and studied . . .
Ancient rubbish tips
ROMAN BAKED BEANS

But all this was nothing compared to his life-long love affair – with insects. The barmy baron devised an awesome ant experiment...

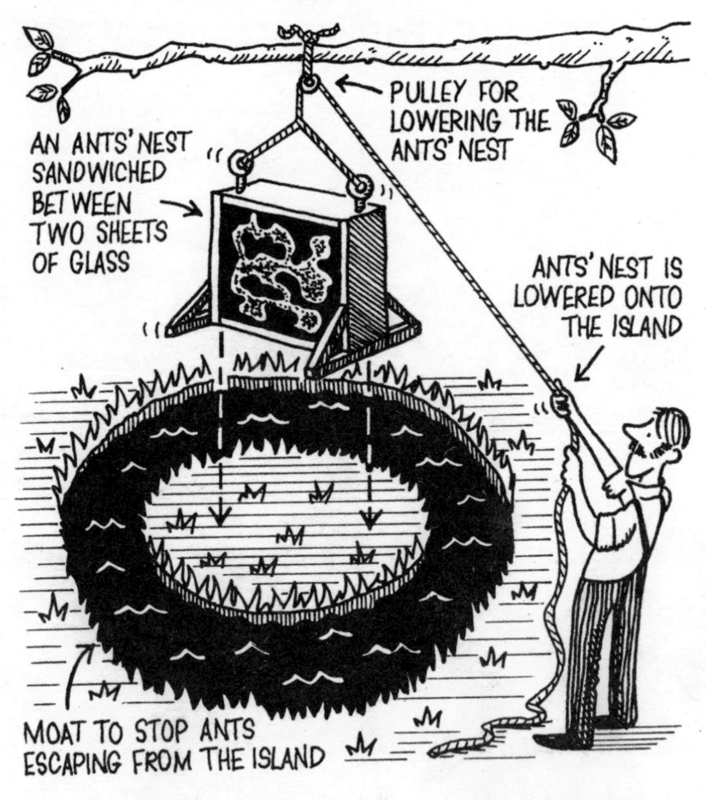

...and discovered...

1 Ants can be ancient. Worker ants can live for seven years, and queen ants for 14 years before they die of old age.

2 Ants respond strangely to sounds – they listen through their legs!

3 Tiny ugly bugs hide in ants' nests.

He devised another ant experiment ... with mazes, obstacle courses and a table with movable rings – all ant-sized, of course. He wanted to find out if ants had a sense of direction. What do you think he discovered?

a) We're talking ant brains here – the ants all got lost.

b) Ants are a bit like sheep – they always follow the ant in front.

c) Ants are really bright. They can judge directions using the sun's rays, even on a cloudy day – so they found their way out.

ANSWERS

a) False. **b)** Partly true. Ants do follow one another – the leading ant makes a trail for the others to follow. **c)** Amazing but true. Ants are better at finding their way than some humans.

ANT AROMAS

Scents are very important to ants. Scientists have discovered several ant scents each of which makes ants do different things. Imagine you were a scientist observing different kinds of ant behaviour. Could you match up the ant behaviour to the smell that causes it? Here are some answers to get you started: **1 c) 2 g) 5 d)**.

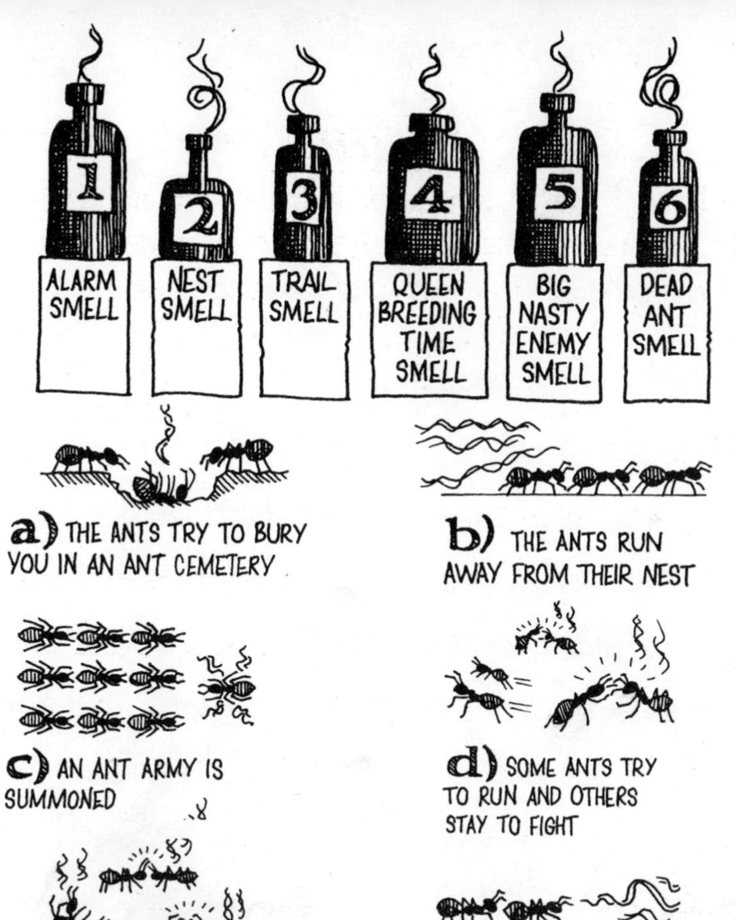

a) THE ANTS TRY TO BURY YOU IN AN ANT CEMETERY

b) THE ANTS RUN AWAY FROM THEIR NEST

c) AN ANT ARMY IS SUMMONED

d) SOME ANTS TRY TO RUN AND OTHERS STAY TO FIGHT

e) ANTS FIGHT EACH OTHER

f) THE ANTS FIND THEIR WAY HOME

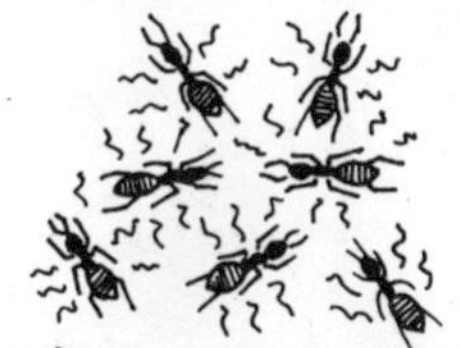

g) THE ANTS DON'T DO ANYTHING IF YOU HAVE THIS SMELL

h) MALE ANTS ARE ATTRACTED BY THIS SMELL

ANSWERS 3 f) 4 h) 6 a)

SLEAZY BEES

Ants and bees belong to the same gruesome group of ugly bugs. So it's no surprise to find some bee species live in nests ruled by queens. Humans tend to say that bees are "good" because they make honey – but bees can be bad in their own horrible way. You'll get a buzz (ha ha) out of teaching your teacher their ugly secrets.

Ugly bug fact file

Name of creature:	Bees and wasps
Where found:	Worldwide. Most bees live on their own. Only a few species live in large nests.
Horrible habits:	They sting people.
Any helpful habits:	Bees make honey and pollinate flowers.
Distinguishing features:	Thin waist between thorax and abdomen. Four transparent wings. Bees have long tongues and often carry yellow lumps of pollen on their hind legs.

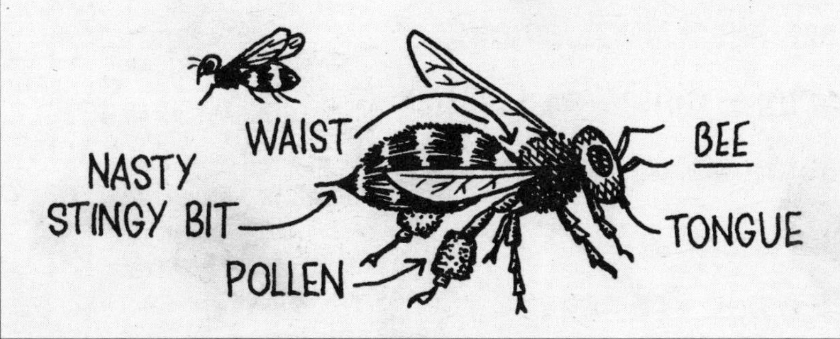

INSIDE THE BEEHIVE

Bees that live together in nests are called "social bees". Well, you'd have to be social to live with *this* lot.

Quarrelling queens Usually there's just one queen in a hive. She spends her time laying eggs. But sometimes more than one queen hatches out, and things can turn rather nasty. The first queen to appear kills off any rivals.

Drowsy drones It's a lovely life for a drone. Your worker sisters keep house for you. And they even feed you. You don't have a sting because you never need to fight anyone. There's just one problem. You've got to battle with hundreds of brothers for a chance to mate. If you mate with a queen you die.

Weary workers What do the workers do? Well (funnily enough) they work. And they work. And they work. In a few short weeks the worn-out workers work themselves to death!

HORRIBLE HONEY

So you love honey. Doesn't the thought of a lovely honey sandwich make your mouth water? And NOTHING is going to put you off it – right? RIGHT. Here's how bees make honey – complete with the horrible details.

1 Bees make honey from the sweet nectar produced by flowers. It's horribly hard work. Some bees collect from 10,000 flowers a day. They often visit up to 64 million flowers to make just 1 kg of honey.

2 That's good news for the flowers because the busy bees also pick up pollen. They even have little leg baskets to carry it. The bee takes the pollen to another flower of the same type. There, some of the precious pollen brushes on to the flower, fertilizes it and so helps it form a seed.

POLLEN BASKETS

3 Why do you think the flower goes to all the bother of making scents, bright colours and nectar. Is it all for us? No! It's to attract bees. Lots of bees means lots of flowers. See?

4 A bee uses her long tongue and a pump in her head to suck up nectar. She stores the nectar in a special stomach.

5 Nectar is mostly water. To get rid of the water, bees sick up the nectar and dry it out on their tongues – ugh.

6 Then they store the honey in honeycomb cells until they need it. That's unless humans steal it for their sandwiches!

BEEWILDER SOME BEES

It's best done on a summer's day in a garden terrace or park where there are lots of bees.

BEES BEE-WARE

Bees have lots of horrible enemies. To stop them, every hive has its guards. The guards don't receive training but if they did it might look like this...

Honey-bee As long as they've got some food you let them in. If not, chase them away! Bee careful. Bees from other hives sometimes steal our honey!

Death's-head hawk moth This nastily-named night raider flies into our hive. It licks our lovely honey with its terribly long tongue. Bee on your guard after dark!

African honey-badger This hairy horror breaks open our honeycomb with its long claws. It makes shocking stinks to drive away our guards. STING ON SIGHT!

Blister-beetle grub Bee careful when you visit flowers. This greedy grub will ambush you! It hitches a free ride to our hive. Then it hides in our cells and guzzles our grubs.

Mouse Another horrible honey hunter. STING TO DEATH! Getting rid of the mouse's body is a bit of a bother. It's too big to move. Cover the body with gooey gum from trees. The gum will mummify the mouse and stop it stinking!

Humans They only want our honey and our bees' wax for polish and candles. Sting them if they get too close. You can't pull your sting from their skin. If you try it'll drag your insides out. Never mind – you'll die a heroine!

Cuckoo bee Don't be a cuckoo and let them in. It's easy to think they're one of us. But once inside they'll lay their ugly eggs.

PRETTY UGLIES

What better on a summer's day than to laze about with a cool drink and watch the butterflies flutter past! And isn't it amazing that there are thousands of different kinds of butterfly throughout the world in an incredible variety of shapes and forms. Pity about their horrible habits and the even more gruesome things they got up to when they were caterpillars!

Ugly bug fact file

Name of creature:	Butterfly
Where found:	Worldwide. The larger butterflies live in tropical countries.
Horrible habits:	Caterpillars eat our vegetables.
Any helpful habits:	Butterflies pollinate flowers and look pretty.
Distinguishing features:	Two pairs of wings, often highly colourful. Narrow body. Long, coiled feeding tube (proboscis) attached to mouth.

THE GOOD, THE BAD AND THE UGLY

The good

1 Butterflies and many moths have amazing coloured patterns on their wings. These colours are made up of tiny overlapping scales and they help male and female butterflies to find each other before mating.

2 Butterflies and moths can detect smells using their antennae. The male Indian moon moth can scent a female over 5 km away. It follows the scent through woods – round trees and across streams – ignoring all other smells. That's like you sniffing your supper 75 km away!

3 Butterflies can smell through their feet! That way they can land on a leaf and know what type it is. Which helps female butterflies to lay their eggs on leaves their caterpillars can happily eat.

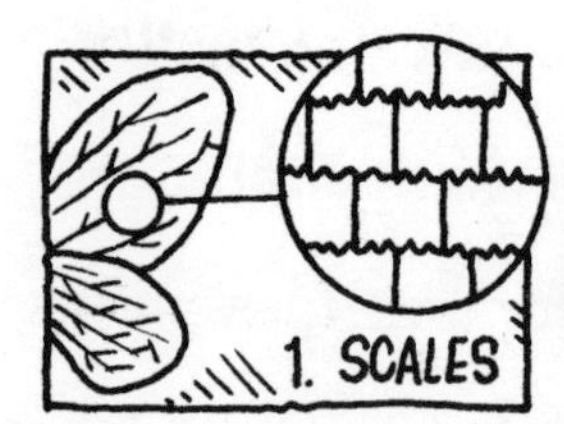

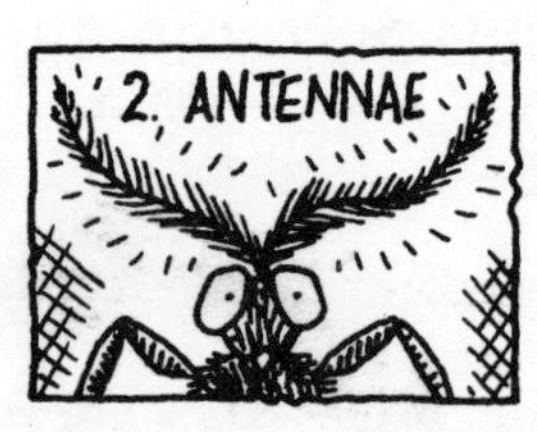

The bad

1 Newly hatched *polyphemous* (polly-fee-mouse) moth caterpillars are tiny. But they start eating straight away and within 48 hours they increase their weight 80,000 times.

That's bad news for the local greenery because the chomping caterpillars can strip all the leaves from a tree.

2 Common large white butterflies really are as common as muck. They fly across the English channel in swarms so vast that a big band of these bad butterflies once stopped a cricket match.

3 But if you're into vast swarms, the African migrant butterfly takes some beating. A scientist once tried to watch a crowd of them fly past. This was a bad idea because the procession continued for three months without stopping!

DAY 6: CAN'T GO ON MUCH LONGER...

And the ugly

1 Ugly scenes have been reported when butterflies get drunk. It's true – the juices from rotting fruit are slightly alcoholic and even one slurp is too much for a butterfly. It flops and droops around on the ground.

2 The gruesome death's-head hawk moth (last seen stealing into beehives) has a sinister skull shape on its thorax. Its equally ugly caterpillars like to nibble the poisonous deadly-nightshade plant. The noxious nightshade makes the caterpillars taste so terrible that no one in their right mind would ever want to eat them.

3 Brown-tail moth caterpillars are also pretty ugly. Their bodies are covered in sharp needle-like hairs that break off in your skin and make it itch like mad.

Could YOU be a large blue butterfly?

The large blue butterfly is – amazingly enough, a large, blue butterfly. In Britain it is very rare and is currently only found in a few places in the West Country. Large blue butterflies are also found in France and Central Europe.

Like all butterflies, the large blue begins life as an egg that hatches into a caterpillar that turns into a chrysalis that turns into a butterfly. But it does horribly odd things on the way. Imagine you were a large blue butterfly. Would you survive?

1 You hatch out. How do you get rid of the remains of your egg?

a) Eat it.

b) Bury it.

c) Throw it at a passing wasp.

2 You live on a wild thyme or marjoram plant. Suddenly your plant is invaded by another large blue caterpillar that starts eating your leaves. What do you do?

a) Agree to share the plant.

b) Eat the rival caterpillar.

c) Hide until it's gone away.

3 After guzzling all the leaves you can, and shedding your skin three times, you fall off your plant. As you amble along, an ant suddenly appears. What do you do?

a) Persuade it to give you a cuddle – then give it some honey in return.

b) Grip its feelers and refuse to let go.

c) Roll over and pretend to be dead.

4 The ant takes you to its nest. It shoves you in a chamber with the ant grubs. What do you do next?

a) Make friends with them.

b) Raid the ants' food supplies and help yourself.

c) Eat the ant grubs.

5 You spend the winter sleeping in the ants' nest. Soon after you wake you hang yourself from the ceiling and turn into a chrysalis. About three weeks later you fall on the floor and crawl out of your nasty damp chrysalis. Congratulations – you're now an adult butterfly! But how do you escape from the ants' nest?

a) You have to dig an escape tunnel.

b) You crawl your way out all by yourself.

c) You pretend to be dead and an ant carries you out.

6 Free at last! What's the first thing you do?

a) Find something to eat – a dead ant will do.

b) Find a mate.

c) Dry out your brand-new wet wings.

And then you fly off to enjoy your new life! Make the best of it – you've only got 15 days to live!

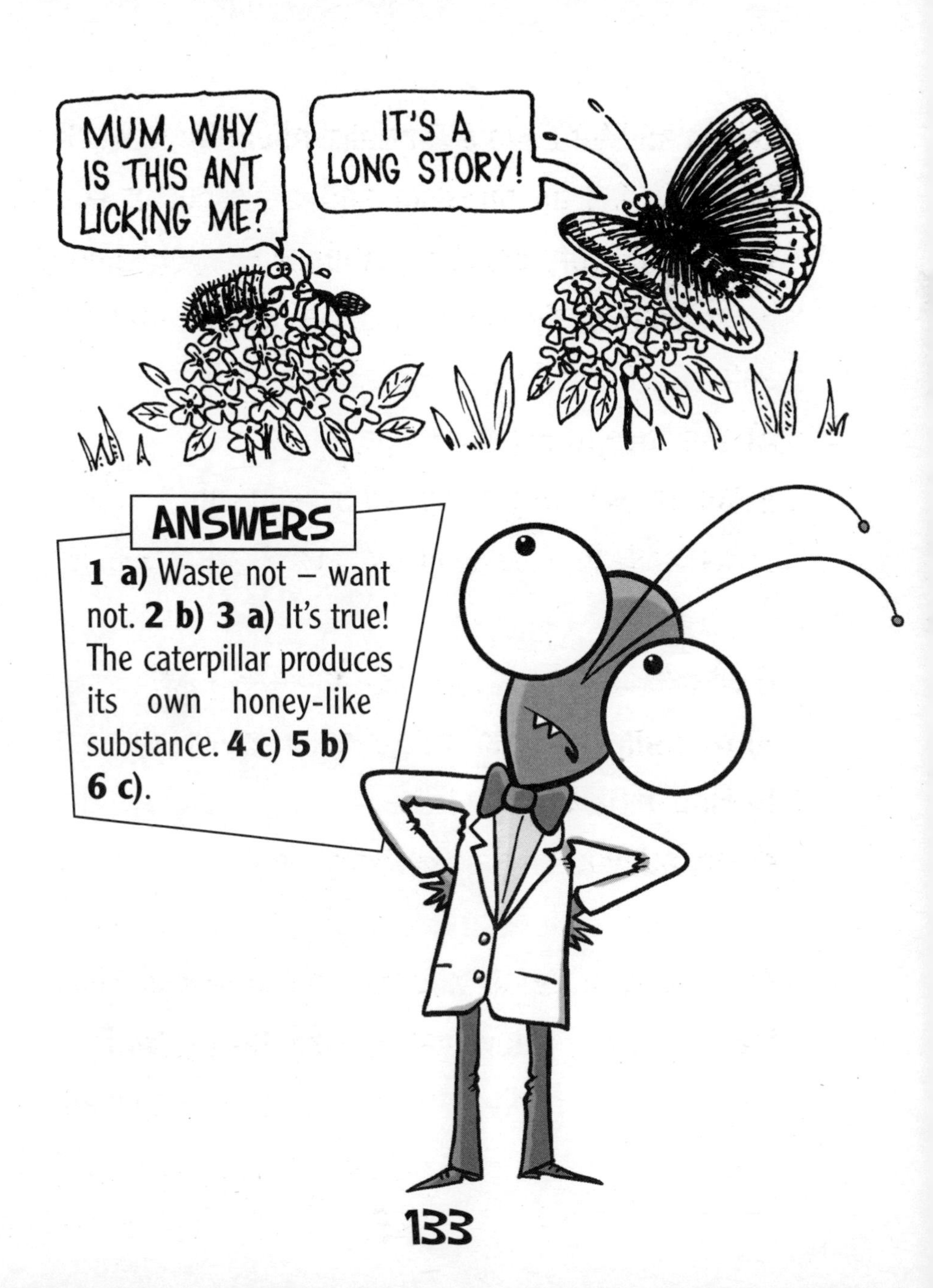

ANSWERS

1 a) Waste not – want not. **2 b) 3 a)** It's true! The caterpillar produces its own honey-like substance. **4 c) 5 b) 6 c)**.

BARMY BELIEFS AND STRANGE SCIENTISTS

For hundreds of years no one knew exactly where caterpillars came from. And there were some pretty strange suggestions. Here's the Roman writer Pliny...

But no one realized that caterpillars were anything to do with butterflies. Then in the seventeenth century, the microscope was invented. All over Europe scientists started to observe insects in gruesome close-up detail.

One of these scientists was Jan Swammerdam (1637–1680) who lived in Holland. As a young man he studied medicine. But he much preferred studying insects to humans! His work was very delicate and he even used tiny scissors that had to be sharpened under a microscope. One day he cut open a cocoon and found ... mixed-up gooey bits of butterfly. Jan had proved that caterpillars turn into butterflies.

But people didn't believe him. The introduction to his insects book, written in 1669, didn't help either. Swammerdam said that the way insects changed their form was...

But as more scientists studied butterflies they discovered that Jan was quite right. These scientists were the first lepidopterists – a horribly complicated name for people who study butterflies and moths.

LETHAL LEPIDOPTERISTS

Nowadays lepidopterists are mild-mannered folk who enjoy observing and photographing butterflies in what is left of their natural surroundings. It wasn't always like that.

1 In the eighteenth century, fashionable ladies wore brightly coloured butterfly and moth wings as jewellery.

2 Traditional butterfly hunters raced after butterflies with big nets shouting, "There she goes!" When they caught an unfortunate flutterer they plunged it into a bottle of poison and pinned it to a board – *horrible*!

3 In the nineteenth century, hunters collected hundreds of butterflies from tropical forests in New Guinea. When the butterflies soared too high they fired guns loaded with fine shot to bring them down!

4 The British collector, James Joicey, spent a fortune over 30 years paying people to collect butterflies for him. By 1927 this millionaire's son had run out of cash. But when Joicey died in 1932 his collection numbered 1,500,000 dead butterflies.

IS YOUR TEACHER A LEPIDOPTERIST?

Find out the easy way with this teacher-teasing test.

1 How can you always tell a moth from a butterfly?
a) Moths come out at night and butterflies in the day.
b) Moths rest with their wings flat. Butterflies rest with their wings upright.
c) Moths don't have knobs on their antennae.

2 How does a hairstreak butterfly avoid having its head bitten off?
a) It has a dummy head.
b) It has a head with armour on it.
c) It bites first.

3 Silk comes from the cocoons spun by the silkworm moth caterpillar. According to legend this was discovered by a Chinese Empress in 2640 BC. But how did she make her discovery?

a) By careful scientific observation.
b) Her cat brought in a cocoon to show her.
c) A cocoon fell into her cup of tea.

4 Where does a cigar-case bearer caterpillar live?
a) In a cigar case.
b) In a little house made of bits of plants joined with silk.
c) In the fur of animals.

5 How can you tell when a butterfly is old?

a) Ragged wings
b) It goes grey.
c) Droopy feelers.

ANSWERS

1 c) All the others are generally true but not always. 2 a) One head is just a decoy. A horrible hunter thinks it's bitten off the butterfly's head. Instead all it's got is a mouthful of wing. 3 c) The hot liquid loosened the strands of silk. 4 b) 5 a) Old butterflies have actually been around for just a few weeks. They hardly ever live much longer.

SAVAGE SPIDERS

The horrible thing about spiders is that you can't get away from them. You can see their webs on plants and washing lines and in garden sheds. And when you come home you'll probably find spiders hiding there too. Spiders aren't insects but that doesn't make them any less horrible. In fact, more people are scared of spiders than are scared of insects. Maybe it's because spiders have some seriously savage habits.

Ugly bug fact file

Name of creature:	Spider
Where found:	Worldwide. On land and in fresh water.
Horrible habits:	Paralyses prey with poison fangs and sucks out the juices.
Any helpful habits:	Keeps down the numbers of insects.
Distinguishing features:	Head and thorax joined. Separate abdomen. Four pairs of jointed legs. Eight eyes. Produces silk. Inside is a breathing organ called a book lung.

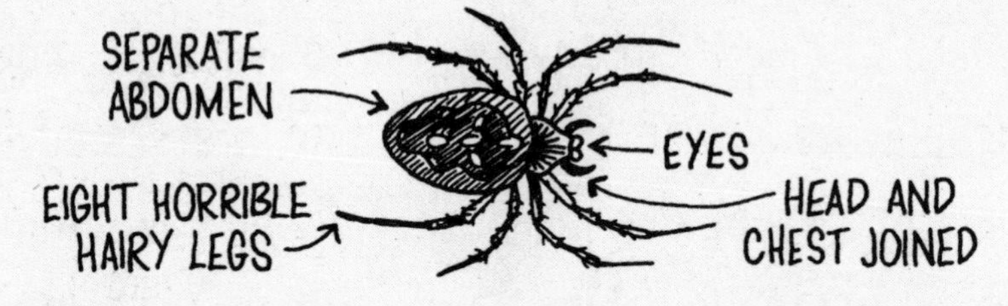

Spiders can't always be savage, surely? They care for their young – sometimes. Mummy wolf-spiders often carry baby spiders on their backs. Ah, how sweet. It's a pity mum eats dad and the babies eat each other. And then there are the really savage bits. Read on at your own risk!

TEACHER'S TERROR TEST

Turn the tables on your teacher as you test his terror tolerance.

1 How do spiders avoid getting caught in their own webs?
a) Nifty footwork.
b) They have oily non-stick feet.
c) They slide down a line and pulley.

2 How long can a spider live?
a) Six months
b) 25 years
c) 75 years

3 When a spider sheds its skin what parts does it get rid of?
a) Its skin.
b) The front of its eyes
c) The lining of its guts and book lung (breathing organ).

4 What does a spider do with its old web?
a) Wear it.
b) Throw it away.
c) Eat it.

5 What does a spitting spider do?
a) It spits a poison that kills its victims as they try to escape.
b) It lassoes its victims with a 10-cm squirt of silk that ties them to the ground.
c) Nothing. It sits around looking strangely sinister.

6 How do small spiders fly through the air?
a) They use electricity in the atmosphere.
b) They inflate their bodies like tiny balloons.
c) They spin little silk parachutes.

7 What, according to legend, is the best way to cure the bite of a tarantula spider?
a) A cup of tea.
b) A lively folk dance.
c) Suck out the venom.

8 How many spiders are there in one square metre of grassland?
a) 27
b) 500
c) 1,795

9 How does a spider get into your bath?
a) It crawls up the drainpipe but can't climb out of the bath.
b) It drops down from the ceiling but can't climb out of the bath.
c) It crawls out of the taps but can't climb out of the bath.

ANSWERS

1 b) 2 b) Tarantulas can live this long. **3** All of these! **4 c) 5 b) 6 c)** Sometimes it's a length of silk and sometimes it's a silken loop that acts just like a parachute. **7 b)** A spider's bite is supposed to make you dance madly – that's called tarantism. The tarantella folk dance is supposed to cure the bite. **8 b)** Scientists reckon there are two billion spiders in England and Wales! **9 b)** The spider drops in for a drink of water. But the sides of the bath are too slippery for the spider to climb out again.

SAVAGE SPIDER FILE

So your teacher's terrified of spiders? Here are a few rational reasons why they're quite right to be scared.

The bird-eating spider - a terrifying tarantula

Description: Big. Can grow to 25cm long including legs.

Lives in: South America

Fearsome features: Scarifyingly hairy

Marital status: Single

Horrible habits: Eats birds and frogs.

The bad news: It has a painful bite.

The very bad news: Those hairs can give you a nasty rash.

The absolutely appalling news: People keep them as pets.

The wandering spider

Description: 12cm leg-span with hairy legs.

Lives in: Brazil

Fearsome features: Said to be the most dangerous spider in the world.

Marital status: No one dare ask.

Horrible habits: Comes into houses uninvited. Wanders around biting people.

Redeeming features: Keeps your home free of bugs and burglars.

The bad news: Its bite is poisonous.

The very bad news: Nasty personality. Likes fighting and frightening. When disturbed, bites first and asks questions later.

The absolutely appalling news: Hides in clothing and shoes. Although an antidote exists, the poison can kill.

The black widow spider

Description: Body 2.5 cm long. Always in black with a sinister red mark on her underside.

Lives in: Southern USA

Fearsome features: One of the most poisonous spiders.

Marital status: Probably a widow

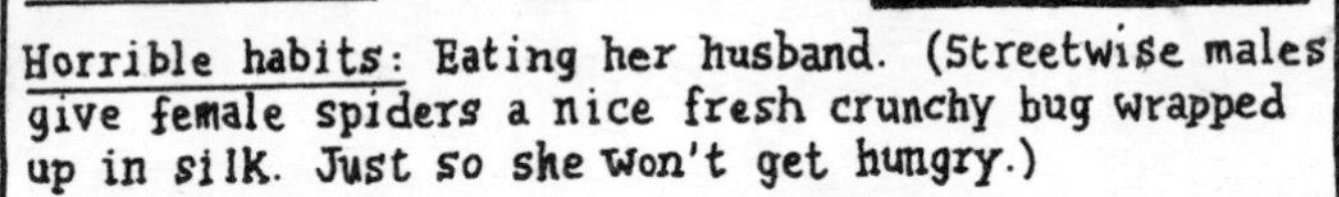

Horrible habits: Eating her husband. (Streetwise males give female spiders a nice fresh crunchy bug wrapped up in silk. Just so she won't get hungry.)

Redeeming features: Rarely bites people. A shy spider who doesn't like fighting and only bites if you come across her unexpectedly.

The bad news: She hides in places where you come across her unexpectedly.

The very bad news: Such as toilet seats.

The absolutely appalling news: And her poison is absolutely deadly. It's said to be 15 times deadlier than a rattlesnake's.

STRANGE SPIDER BELIEFS AND EVEN STRANGER SCIENTISTS

Some spider scientists had strange ideas and others were involved in strange experiments. Spider science (arachnology) started with the

Greeks. But Greek writer, Philostratus, had some rather strange ideas about spiders. He reckoned that spiders spin silk to keep warm. Nice try, Phil. Mind you, the Romans weren't much better. According to Pliny, spiders appeared from seeds that grew in rotting material.

We've all heard the rhyme about little Miss Muffet who got scared away by a great big spider. But did you know that she was a real person? Her name was Patience Mouffet and she was unfortunate enough to be the daughter of strange sixteenth-century spider scientist, Dr Thomas Mouffet. Why unfortunate? Well, her dad used to dose her with live spiders whenever she had a cold. As a special treat Patience got to eat mashed up spiders on toast.

Brave Baerg Dr W J Baerg of Arkansas, USA, conducted some strange experiments with the aim of discovering exactly how deadly a venomous spider bite really was. In 1922 he deliberately allowed himself to be bitten by a poisonous black widow spider! The first test was a failure – the spider wouldn't bite. So Dr B tried again and this time he was delighted to get a nasty nip. When he got out of hospital three days later the spider scientist recorded that he'd felt unbearable pain. Now, there's a surprise.

In 1958 Dr Baerg was at it again. This time he decided to test spider bites on guinea pigs and rats rather than himself. But the intrepid investigator didn't escape pain altogether. He had an unhappy accident with a Trinidadean tarantula. Baerg was lining up the hairy horror to bite an unfortunate white rat when it bit him on the finger. (That's the spider not the rat.) Luckily Dr B found that the poison didn't harm him. So he allowed himself to get bitten by a Panamanian tarantula instead and this time suffered a stiff finger. So brave Dr Baerg concluded that tarantula bites weren't so bad after all!

Could YOU be a spider scientist?

Can you predict the result of this strange spider experiment?

In 1948 Professor Hans Peters noticed that garden spiders always spun their webs at 4 o'clock in the morning. So he fed some spiders with caffeine (that's the chemical in coffee that wakes people up) and others with sleeping pills to see what would happen. What do you think he discovered?

a) Spiders are affected just like us. The spiders stimulated with coffee woke up at 1.30 am and then worked all night. The spiders drugged with sleeping pills slept until 10.35 am.

b) Spiders are completely different from humans. The caffeine-stimulated spiders went to sleep. And the spiders drugged with sleeping pills worked harder than ever.

c) The urge to spin webs was stronger than any drug. The spiders made some odd-looking webs. But they continued to start work at 4 am.

ANSWER c).

BET YOU NEVER KNEW!

Spiders have spun webs in space. On 28 July 1973 garden spiders Arabella and Anita boldly blasted off into space to visit the Skylab space station. Their mission – an experiment to find whether they could spin webs in zero gravity. Their first efforts were untidy. They weren't used to floating around weightless. Later efforts were more successful although poor Anita died in orbit.

WEIRD WEBS

Spiders spin silk to produce their intricate webs. The webs they make catch flies and other unlucky creatures. But the more you find out about webs the weirder they seem.

1 To make one web, spiders need to spin different types of silk.

- Dry silk a thousandth of a millimetre thick for the spokes of a web.
- Stretchy silk covered in gluey droplets for the rest. The sticky bits take in moisture and stop the web drying out.
- Other kinds of silk for wrapping up eggs and dead insects.

2 Webs come in many shapes and sizes. Have you ever seen any of these?

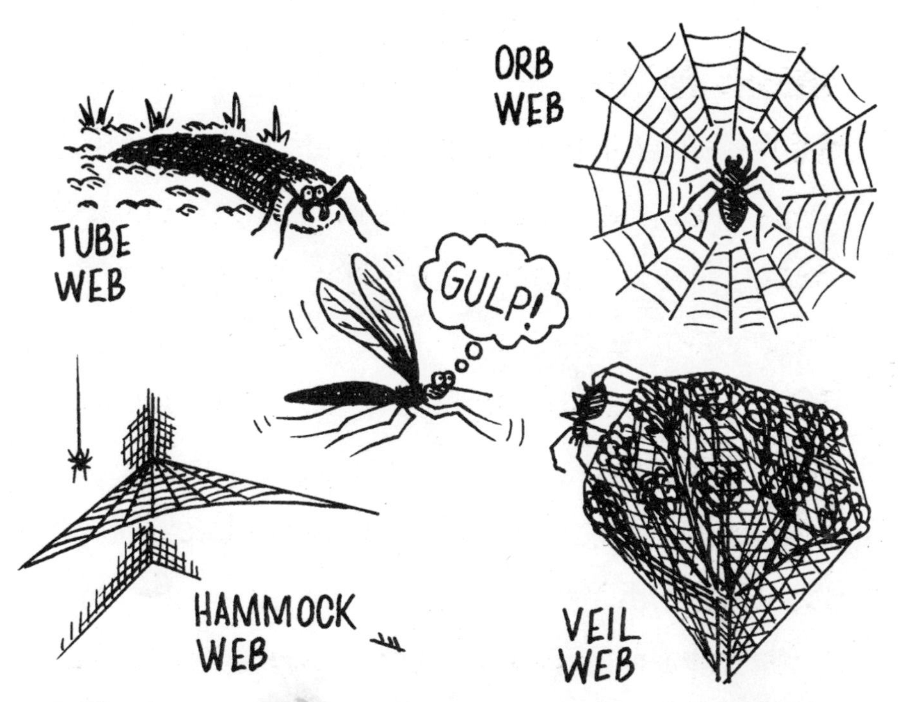

3 The house spider makes a hammock-shaped web. The spider spits out bits of insect and leaves them lying around for someone else to tidy up – a horrible habit!

4 The trap-door spider digs a tunnel with a trap door at one end. The spider waits within. It grabs a passing insect and pulls it down. The door closes and the innocent victim is never seen again.

5 The purse-web spider makes a purse-shaped web. This savage spider then stabs its victims through the web with its long poison fangs. Then before settling down to dinner it carefully repairs the tear.

6 Sinister *nephila* (Nef-illa) spiders spin giant webs up to 2 metres across to catch insects and sometimes even birds. And even fish aren't safe – in the early part of this century people in New Guinea used the silk to make fishing nets!

7 The web-throwing spider chucks its web over insects as they work under its hiding place. Then the spider drops in for tea.

8 When an insect gets caught in a spider's web, it struggles and the vibrations alert the spider. But the devious ero spider manages to sneak on to its enemy's web and bite the spider before it even realizes its got a visitor. Evil ero then sucks its victim dry and scurries away, leaving an empty spider husk sitting in its web!

9 In some places in California, spiders' webs fall like snow. The flakes are made up of mixed-up spiders' webs blown together by the wind.

BET YOU NEVER KNEW

It's actually possible to spin and then weave spider silk. In 1709 Xavier Saint-Hilaire Bon of Montpelier showed the French Academy several pairs of spider-silk gloves. But according to a scientist who investigated the topic it takes 27,648 female spiders to make less than half a kilogram of silk. That didn't stop more people trying to spin spider silk. In the early 1990s the US Defence Department looked into the possibility. The reason – the silk is light, strong and very springy. Ideal for making bulletproof vests!

And now you know all about these hairy horrors...

DARE YOU MAKE FRIENDS WITH ... A SPIDER?

To save spinning your own silk, try asking a spider to make some for you.

1 Cut a plastic lemonade bottle in half. (Get an adult to help you.)

2 Add soil and twigs to the bottom half.

3 Now find your spider. Sheds and out-houses are good places to look. If you find a web the spider is normally hiding nearby. One spider is enough. Add two and one will eat the other! Be gentle, though – spiders are easily hurt!

4 Tape up the two halves of the bottle.

5 Feed your new friend with a small fly through the top of the bottle.

6 Check to see if she's spun any silk or made a web. If she has, go ahead and try knitting yourself a nice pair of spider silk gloves.

For us humans the most horrible thing about insects is the way they bite us. And suck blood and sometimes give us horrible diseases too. Maybe that's why people call a crowd of insects a "plague". Plagues are deadly diseases. In the past 10,000 years more people have died from diseases carried by insects than any other single cause. Help!

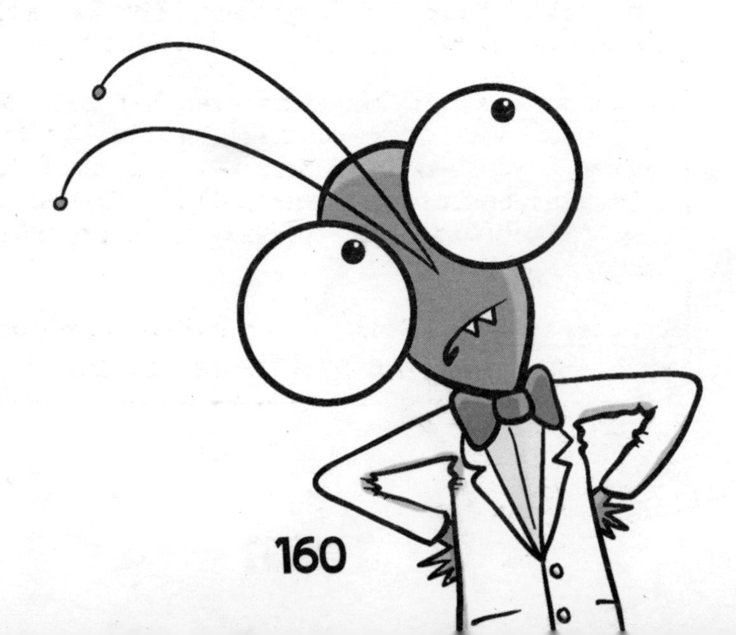

ROGUES' GALLERY

Here are the chief culprits.

Malarial mosquito

<u>Sex:</u> Female

<u>Habits:</u> Sucks blood before laying her eggs, while Mr Mosquito prefers plant juices.

<u>Weapons:</u> A long snout for sticking into people and a pump-action saliva gun to stop your blood setting.

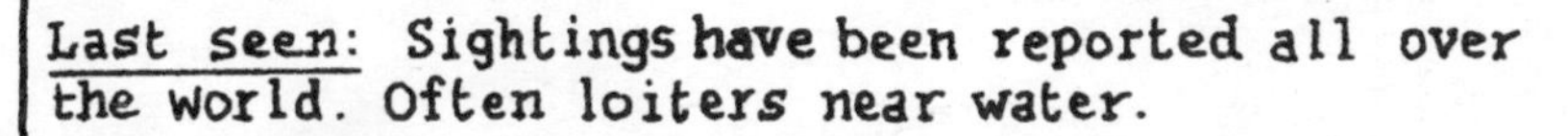

<u>Last seen:</u> Sightings have been reported all over the world. Often loiters near water.

<u>Known crimes:</u> In hot countries her bite passes on germs that cause malaria. Victims suffer raging fever and feel very hot and then very cold. Responsible for one million deaths each year. Sometimes gives us yellow fever into the bargain.

<u>Danger rating:</u> Beware! Two billion people live in areas threatened by this brutal blood-sucker.

Body louse

Description: 1.5-3.5mm long. Has no wings.

Weapon: Blood-sucking tube

Habits: Sucks blood

Last seen: Hiding in the seams of clothes.

Known crimes: Disgusting droppings can contain germs that cause the deadly disease, typhus. Victims scratch the droppings and grisly germs into their skin.

Danger rating: Nasty. But nothing that a good bath and clean clothes can't cure.

Known associates: Head lice, or "nits", live in hair. They quite like clean hair and happily hop from head to Head. (Yes. Head teachers can get them too!)

Tsetse fly

Habits: Sucks blood. Known to drink up to three times its own weight in a single sitting. Likes a challenge - enjoys biting through rhinoceros skin.

Last seen: Many parts of Africa.

Known crimes: Its bite passes on germs that cause sleeping sickness. This deadly disease causes fever, tiredness and death.

Danger rating: In Africa 50 million people are at risk plus countless cattle, camels, mules, horses, donkeys, pigs, goats, sheep, etc. etc. etc.

Benchuca *(ben-chooka)* **bug**

Last seen: South America

Habits: Creeps up on you at night and stabs you with its pointed snout. Sucks a bit of blood and scarpers before you squash it.

Known crimes: Spreads Chagas' disease. Result – tired and feverish humans.

It took humans many years of painstaking research to track down the culprits of these terrible diseases and to decide what to do about them.

MALARIA MYSTERY SOLVED

In the nineteenth century Scottish born scientist Patrick Manson discovered how mosquitoes pass on malaria. Here's how he did it.

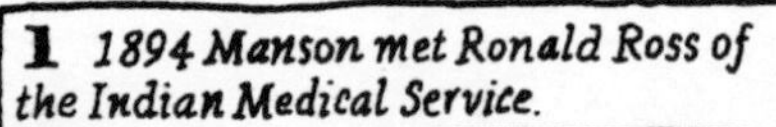

4 1900. Manson wanted more proof. He sent two assistants off to spend three months living in a swamp full of mosquitoes.

Is this really such a good idea?

6 Now for the ultimate test! Manson allowed his own son and another volunteer to be bitten by a mosquito that carried the parasite.

Yummy!

OW!

PLAGUE PUZZLE

Some diseases were even more puzzling than malaria. Can you piece together the gruesome clues to unravel the cause of the deadly bubonic plague?

1346 It came from the East and in the next six years 25 million people died. Peasants died in their fields, and in England three Archbishops of Canterbury died in a single year. People lived in terror of THE BLACK DEATH.

1855 The plague ravaged China. In 1894 it hit the Chinese ports and in Hong Kong the death toll soared. The harbour was crammed with steam ships. And these ships took the disease to Japan, Australia,

South Africa and the Americas. The plague reached India and killed six million people in ten years.

1898 In Bombay. Dr Paul-Louis Simond of the Institut Pasteur was a worried man. The fearless French doctor had been sent to India to find the cause of the plague. Day and night he wrestled with the same fiendish puzzle. In the stricken city thousands of people were dying. All developed fist-sized bulges under their armpits followed by fever and death. But how and why?

Day after day Simond scoured the squalid streets in search of an answer. Everywhere he noticed dead rats – 75 in one house. It was extremely unusual to find so many dead rats all together in one place.

They must have died quite quickly but what had killed them? And why was it that any humans who touched the rats seemed to fall sick with the plague? These plague rats seemed to have more fleas than healthy rats. And the fleas bit people, too.

The monsoon rain buffeted the outside of the makeshift lab in a tent. Inside, Simond risked his own health as he cut up the dead rats. Then he made a dramatic discovery. In the rat's blood he found the germs known to cause the plague.

But what was the cruel connection between rats, fleas and humans? At long last the answer came. The intrepid scientist had solved the most terrifying mystery of all time. That evening he wrote in his diary in a frenzy of excitement.

But what was that crucial connection?

a) A flea bites a rat and passes on the plague. The rat bites a human and passes on the plague.

b) A flea gets plague from biting an infected rat. The flea bites a human and passes on the plague.

c) A human gets plague from an infected flea bite. The plague-crazed human bites a rat and passed on the plague.

ANSWER

b) Microbes multiply in the flea's gut until it can't feed. The hungry flea bites a human and injects millions of germs.

Although Simond had the answer it took another 20 years before scientists accepted that he was right. It wasn't until 1914 that they fully understood the effects of the plague on fleas. Already vaccines were being developed against the plague and these together with insecticides and rat poisons have reduced the danger of plague epidemics in the future.

BEAT THE BUGS

Hopefully you can avoid getting a horrible disease from a biting bug. But it's hard to avoid getting bitten. Here are some danger zones.

1 Bed During the day bed bugs hide in cracks and behind wallpaper. Then at night out they pop for a midnight feast of blood.

2 Riverbanks Blackflies launch dawn and dusk raids.

3 Fields at dawn Ticks can lurk in the long grass. They prefer dogs for dinner but if one isn't handy they'll make do with you.

4 Bogs and marshes Millions of midges fly around seeking blood for breakfast. Close up they're too small to see clearly. You can't see their wings because in some varieties they beat at an awesome 62,760 strokes a minute. That's why some people call midges, "no-seeums". But you know-um when they bite you.

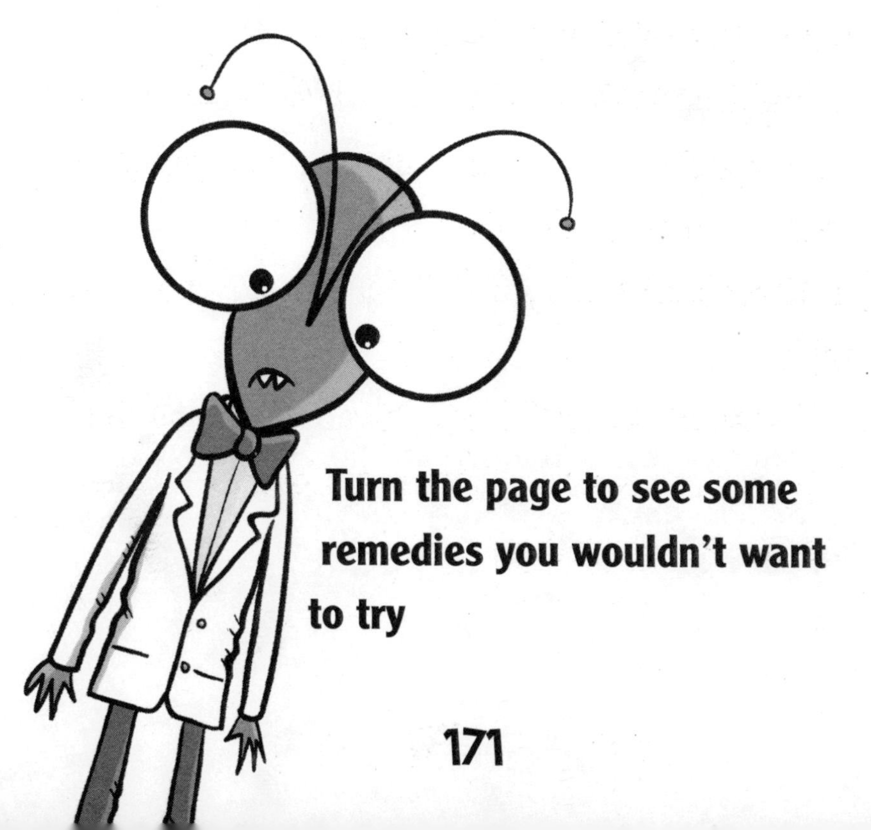

Turn the page to see some remedies you wouldn't want to try

1 Tsetse fly trap no. 1

Remedy: Keep a pet ox.

Notes: Scientists have found that the terrible tsetse is attracted to smelly ox breath. In Zimbabwe similar smelling chemicals lured thousands of tsetses into poisoned cloth traps.

Drawbacks: Smelly ox breath. Feeding your ox. Having to take it to school with you.

2 Tsetse fly trap no. 2

Remedy: Ferment some cassava.

Notes: In Zaire people make beer from cassava roots. The messy mixture produces carbon-dioxide gas that lures the flies to their doom.

Drawbacks: People might start drinking your beer. This could cause embarrassing situations – especially at school.

3 Bed bug beaters

Remedy: Let loose an army of Pharaoh ants in your bedroom.

Notes: Pharoah ants eat bed bugs.

Drawbacks: How to get rid of the Pharaoh ants. Try smearing your duvet with jam. Or borrow an anteater.

4 Biting bug barbecue

Remedy: Light a really smoky bonfire.

Notes: Most biting bugs don't like smoke.

Drawbacks: Not a very sensible thing to do. Ever. Especially not recommended inside the home or at school.

5 Flea fighters

Remedy: Use flea mites to fight fleas.

Notes: Tiny mites infest fleas in the same way that fleas infest people. All you need to do is to capture a flea and add some mites. (You need a microscope and a steady hand to do this.)

Drawbacks: Doesn't get rid of fleas. But it does give them a taste of their own medicine.

6 Pest poisoner

Remedy: Squirt biting bugs with DDT.

Notes: In the 1940s this insecticide was used to rid the southern USA and parts of Africa and South America of malarial mosquitoes.

Drawbacks: By 1950 two types of mosquito were immune to the powerful poison. Worse still, DDT harmed bug-eating animals and the animals that ate those animals. Note- humans still spend millions of pounds each year inventing new kinds of insecticide.

A remedy you might like to try

Some aromatic plant oils ward off insects. You can buy these oils from herbalists or natural cosmetics shops. You could try using citronella oil on a warm summer's evening...

1 Drip a few drops of the oil on a damp piece of cotton wool.

2 Put the cotton wool in a warm place indoors.

3 When the room is full of scent open the window and dare any biting bugs to come in!

BET YOU NEVER KNEW!

Long before humans discovered camphor, female assassin bugs were rubbing their abdomens in camphor resin. When the females lay eggs, the eggs get coated in camphor goo. The powerful pong keeps other bugs at bay.

As if escaping from spiders, fish, lizards, frogs, toads, small mammals and even horrible humans wasn't enough, insects seem to spend most of their time playing hide and seek with each other. And they don't just do it for fun. Insects have to eat, after all, and they don't want to get eaten. So they use some cruel and cunning tricks to get one up on their ugly bug enemies.

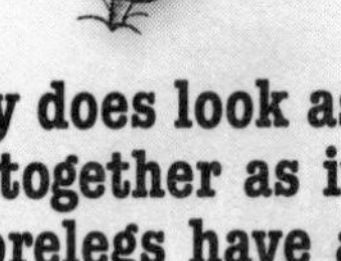

BET YOU NEVER KNEW!

Horrible hunter, the praying mantis really does look as though it's praying. It holds its forelegs together as it waits for a tasty snack to pass by. Its forelegs have a jagged edge, just like a saw blade. The praying mantis will catch and skewer its bug lunch in just a twentieth of a second – then bite off its ugly little head!

INSECT SURVIVAL

If you were an insect, would you stay alive? Try this crash course in survival skills to help you decide.

Tactic number 1: Pretend to be something else
You've obviously got an advantage if you already look like something else – and quite a few insects do. Which of these ordinary objects might really be insects?

a) A leaf
b) A sweet wrapper
c) A twig
d) A stick
e) A thorn
f) A bird dropping

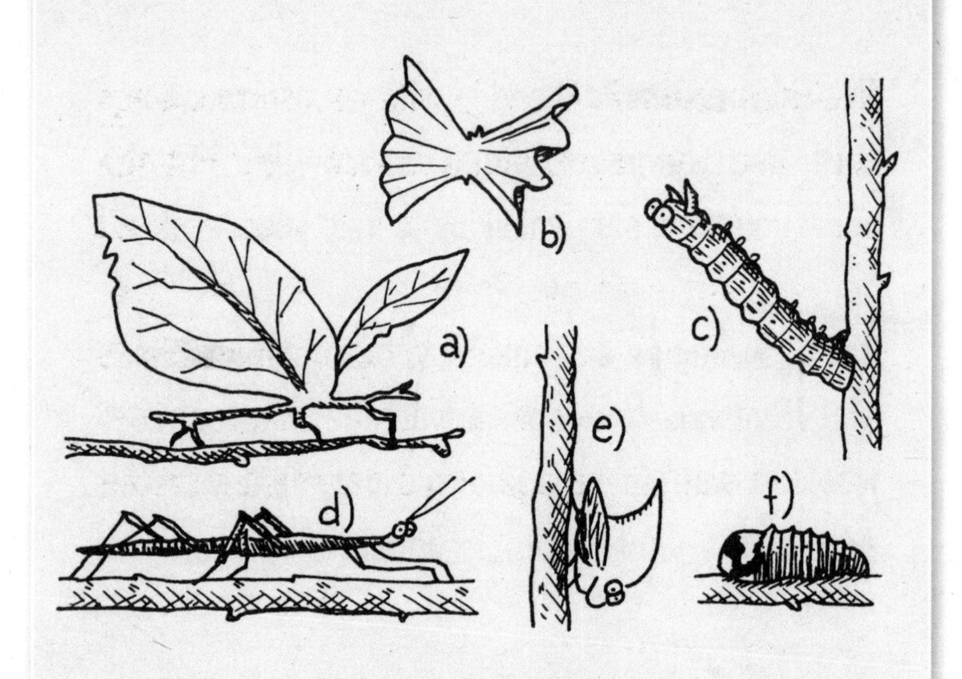

ANSWERS

All except **b)** could be insects. See for yourself: **a)** is a Japanese leaf insect, **c)** is a European swallow-tail moth caterpillar, **d)** is of course, a stick insect, **e)** is a treehopper, **f)** is a hairstreak butterfly caterpillar.

Tactic number 2: Blend in with your surroundings
Look like your surroundings, stay still, and the hunter might just miss you. The clear-winged butterfly, for example, is invisible – it has see-through wings that make it almost impossible to spot. But you may not be such a lucky bug. You wouldn't want to be a poor old peppered moth for example...

The problems of the peppered moth

This light speckled moth likes to hang around on light speckled trees. Perfect – not a horrible hunter in sight. Then came industry and pollution. And all the trees turned black.

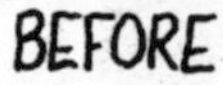

Suddenly the moths stood out like sore thumbs. The birds had a bonanza munching millions of moths. But some moths survived – only the ones with very dark colouring, though.

For years these dark moths had had it tough, trying not to be noticed on all those light speckled trees. Suddenly they had a bright future hiding on dark sooty ones. Or they did – until the cities started getting cleaner and the trees started getting lighter again!

Tactic number 3: Brilliant bluffs

Disguise yourself as a dangerous character and you can bluff your way out of danger.

1 Hover-flies are harmless little things. An ideal dish

for an ugly bug's dinner. Or they would be if they weren't wickedly disguised as wasps! Clear-wing moths try the same trick too, but they're even better bluffers – they can make the sound effects, too!

2 Ladybirds taste terrible. On the other hand, fungus beetles taste quite nice (if you're another insect, that is). That's why fungus beetles go around pretending to be ladybirds.

3 Some butterflies even disguise themselves as other types of butterfly. In South America there are four strangely similar-looking varieties of butterfly. Only one is nasty to eat. The other three are just plain copy-cats.

4 Another great little bluffer is the hawk moth caterpillar. Its head looks fairly normal, for a caterpillar – but its rear end looks more like a snake's head!

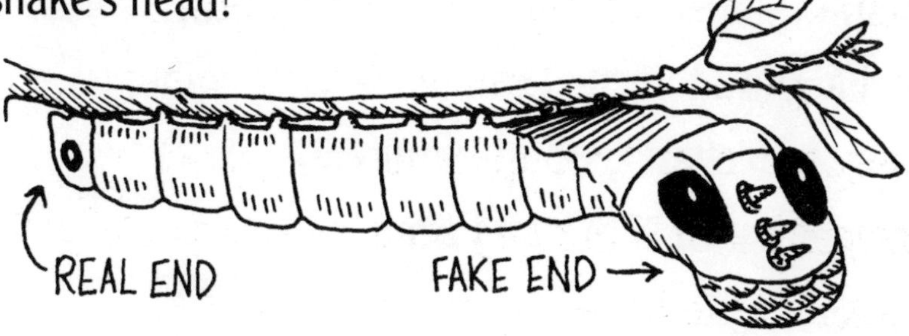

But, beware of disguises...

5 The African dead-leaf cricket is cleverly disguised to look just like any rotten old leaf. It's got one little problem. The casque-headed frog also looks like a rotten old leaf – and this rotten old leaf likes nothing better than a cricket for its tea.

6 Best bluffer of all has to be the puss moth caterpillar.

Take a look at its ugly mug! Would you want to meet that on a dark and stormy night? You'd do well to avoid this little cruel cat, as it can spit out its half-digested dinner mixed with awful acid.

Tactic number 4: Horrible hiding places

One sure way to avoid being eaten is to hide somewhere horrible. That way no one can find you and no one would want to either!

For example, the plume-moth caterpillar hides out in the sundew plant. The sundew plant eats flying insects, but the caterpillar is safe inside the plant and it gets to drink the droplets of sundew goo and munch on the sundew's insect tea.

Froghoppers hide inside a mass of foam. It looks a bit like bubble bath, but they make it themselves. The foam stops froghoppers from drying out in the sun, and it has a revolting flavour to put off horrible hunters.

DARE YOU MAKE FRIENDS WITH ...

A FROGHOPPER?

1 Look for its little drops of foam on the long grass in early spring. The foam is sometimes called "cuckoo spit" – you can guess why!

2 Gently brush the foam away and you'll see a little greenish insect hiding underneath.

3 Watch carefully as it blows bubbles from the end of its body to cover itself up again. The froghopper sucks plant juices and mixes them up with its own natural bubblemaker to make the froth.

OK, so maybe the froghopper doesn't want to make friends, but you've got to admit it's got a devilish disguise.

UGLY BUGS VS HORRIBLE HUMANS

Since the day that a caveman or cavewoman first squashed a cockroach there has been a non-stop war between ugly bugs and humans. It's the biggest war the world has ever known.

You might think that humans have an advantage over insects. A human is far bigger than the biggest insect. So humans can easily squash the insect. Humans are more intelligent than insects. (Well, *most* humans are!) But if you look at what humans and insects can do for their size the picture is very different.

UGLY BUG OLYMPICS

Running winner: One species of cockroach can run 50 times its body length in one second. **Loser:** The fastest human to run 50 times his own body length (about 80 metres) was about nine times slower.

The high jump winner: Fleas can jump 30 cm – that's 130 times their own height. **Loser:** To match that a human would have to jump 250 metres into the air!

The long jump winner: Jumping spiders can leap 40 times their body length. **Runner-up:** Grasshoppers can leap 20 times their own body length. **Loser:** To match that a human would have to leap the length of nine London buses in a single jump!

HUMAN INSECT

Weight-lifting winner: Scarab beetles can lift weights 850 times heavier than their own bodies. **Loser:** To equal that a human would have to lift eight London buses at the same time!

HUMAN INSECT

Walking on the ceiling winner: Flies. **Loser:** Humans can't do this at all.

Surely though, we humans are better at some things. Like building, for example. I mean – there's the pyramids and St Paul's Cathedral and the Taj Mahal. Ugly Bugs can't match that ... can they?

BET YOU NEVER KNEW!

Termites build gigantic nests. One nest contained 11,750 tonnes of sand. The termites had piled it up grain-by-grain and stuck it all together with spit! Beat that – humans!

But who are the dirtiest, the greediest and the most destructive creatures on the planet – ugly bugs or horrible humans? You might find it difficult to choose between them.

FILTHY FLIES

They never give up. It doesn't matter how many times you let them out the window they always come back.

1 Blowflies enjoy eating rotting meat and animal droppings. They lay eggs on rotting meat and even do terrible things to your Sunday roast.

2 The common housefly has common table manners. It drops in for dinner uninvited and sicks up over its food. And then it's been known to serve up a free selection of over 30 deadly diseases.

HORRIBLE HUMANS

Humans are also very persistent. Once they decide to do something they will do it even if it costs the Earth – literally.

1 Humans are the only animals that deliberately destroy their environment. Every second, humans devastate a area of forests, grasslands or swamps to build things for themselves. Each year humans burn an area of rain forest the size of Great Britain.

2 Humans also pollute the world with litter and dangerous chemicals. Every day humans dump millions of tonnes of rubbish into the sea.

3 Human beings are killers. Every hour of the day human destruction and pollution of the environment wipes out an entire species of living plant or animal.

HORRIBLE HUMANS HIT BACK

Day after day humans wage war against insects with every weapon at their disposal. But they've also discovered some surprisingly horrible uses for insects and other ugly bugs.

REVOLTING RECIPES

If you can't dispose of ugly bugs you could always eat them. That's what millions of apparently sane people do throughout the world. Would you want to try any of these dishes?

Starters

Fried and salted termites

An African treat. Tastes like fried pork rind, peanuts and potato chips all mixed up!

L'escargots

Oui, mes amis! The traditional French delicacy. (Snails to you.) Fed on lettuce. Boiled and cooked with garlic, butter, shallots, salt, pepper and lemon juice. Served with parsley. Bon appetit!

Fried witchetty grub

A native Australian delicacy – these are giant wood-moth grubs. They look a bit like fusilli pasta and swell up when fried. Delicious!

Main courses

Stir-fried silkworm pupae

This tasty traditional Chinese dish is prepared with garlic, ginger, pepper and soy sauce. Wonderful warm nutty custard flavour. You spit out the shells. Very good for high blood pressure.

Roast longhorn timber beetle

Deliciously crunchy balsawood flavour. As cooked by the native people of South America.

Fried Moroccan grasshopper

Boiled bug bodies prepared with pepper, salt and chopped parsley then fried in batter with a little vinegar. You can also eat them raw.

Blue-legged tarantula

A popular spider dish in Laos in South-east Asia. Freshly toasted and served with salt or chillies. Flavour similar to the marrow in chicken bones.

Sweets

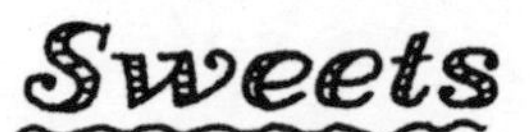

Mexican honeypot ants

A sweet sticky treat.

Baked bee and wasp grubs

An old recipe from Somerset in England. Juicy grubs baked in hot sticky honeycomb.

After your meal

Try one of our tarantula-fang toothpicks as used by the Piaroa people of Venezuela.

UGLY BUGS VS HORRIBLE HUMANS: THE DEBATE

For every argument there are two points of view. And this is certainly true for ugly bugs and humans. See for yourself. Who do you sympathize with most – ugly bugs or humans?

Human point of view	Ugly bug point of view
Ugly bugs sting and bite us.	*Humans trap us, poison us and experiment on us.*
Ugly bugs eat our crops.	*Humans destroy our food plants and plant their crops too close together so we've got nothing else to eat.*

Ugly bugs creep into our homes.	*Humans destroy <u>our</u> homes.*
Ugly bugs spread diseases.	*Humans spread pollution and rubbish.*
Ugly bugs destroy our furniture.	*To us it's only wood.*
Ugly bugs cost us money.	*Who cares about money?*
They destroy our property.	*Who cares about property?*

Ugly bugs just want the same things we want. Nice food and somewhere to live. The problem only comes when their idea of nice food is *your* nice food, and their idea of somewhere to live is *your* bedroom.

THE UGLY TRUTH

You might think that humans are the deadliest enemies of ugly bugs. Wrong. The deadliest enemies of ugly bugs are other ugly bugs. Without ladybirds we'd be overrun by aphids. Without spiders we'd be fighting off flies.

The best way to remove an ugly bug is to get another ugly bug to do the job. When the cottony cushion scale insect invaded California it finished off entire fruit crops. Until humans brought in a type of ladybird to crush the cottony crooks.

And remember all those scary statistics about insects having millions of offspring? You'll be reassured to know that the weight of insects eaten by spiders in a year is greater than the combined weight of all the people on earth. And if ugly bugs really were our enemies do you think we'd stand a chance? Nope. Quite apart from the fact that there's a million of them to every one of us they can do horribly ugly things that we wouldn't even want to dream about.

But there's another side to ugly bugs. All ugly bugs are horribly incredible. Horribly interesting.

And amazingly enough some ugly bugs are even horribly useful to humans.

We rely on ugly bugs to make plants fruit and to eat up rotten plant rubbish. Without insects we'd have no honey and no firefly lanterns. No silk, no jewel beetles and no beautiful butterflies. Admittedly we wouldn't have plagues and half-nibbled vegetables either. Ugly bugs make the world a worse place. But they make the world a better place too. And that's the ugly truth!

HORRIBLE INDEX